FUCHSIAS FOR HOUSE
AND GARDEN

FUCHSIAS

For House and Garden

SIDNEY CLAPHAM

DAVID & CHARLES
Newton Abbot   London

UNIVERSE BOOKS
New York

British Library Cataloguing in Publication Data

Clapham, Sidney
  Fuchsias for house and garden.
  I. Title   1. Fuchsia
  635.9′ 3344   SB412.F8

  ISBN 0-7153-8217-9

© Sidney Clapham 1982

First published 1982
Second impression 1983
Third impression 1983

Typeset by Typesetters (Birmingham) Limited
and printed in Great Britain
by Redwood Burn Limited, Trowbridge, Wilts
for David & Charles (Publishers) Limited
Brunel House   Newton Abbot   Devon

First published in the United States of America in 1982 by
Universe Books, 381 Park Avenue South, New York, N.Y. 10016

82 83 84 85 86/10 9 8 7 6 5 4 3 2 1

Printed in Great Britain

Library of Congress Cataloging in Publication Data

Clapham, Sidney.
  Fuchsias for home and garden.

  Includes index.
  1. Fuchsia. I. Title.
  SB413.F8C58 1982   635.9′ 3344   82-8401
  ISBN 0-87663-404-8   AACR2

Contents

List of Plates

Colour

Black and white

All black and white photographs by J. E. Downward unless otherwise stated

1

The Near-Perfect Garden Plant

Few plants have as much to offer as the fuchsia. It is as near to being the ideal garden or balcony plant as we are ever likely to get. It develops quickly, so that cuttings taken in the spring will make good flowering plants by the summer; it is easy to propagate and grow; it has a long flowering period, in many cases through the greater part of the summer and into the autumn; it can be trained in all sorts of ways; it is as free of pests and diseases as most greenhouse plants—in fact it suffers from none of the virus diseases that trouble so many others; and above all, perhaps, its flowers have a variety of shape and colour scarcely exceeded by those of any other plant. Its only failing is that in many of its finest forms it is not hardy enough to grow permanently outside in the British Isles. In the warmer parts of North America, on the other hand, and particularly in California—where many of the finest new varieties come from—it may be safely grown outside all the year round.

With all its good qualities it is obviously a plant well worth including in your range, not only for people with a greenhouse but also for anyone who has a garden large or small, or even a window box or other plant container. The fuchsia was indeed one of the most popular greenhouse and indoor plants in the last century, but it went out of favour after the first world war, and not until after the second world war did it really take the public fancy again. It is now realised that many varieties (I use this familiar term in place of the more cumbersome but internationally correct term 'cultivars') will make a fine show outside in the summer, and a much smaller number can be grown permanently out of doors. Some of the showier and more delicate varieties can of course be grown to perfection only under glass, in Britain, but it is easy enough to suit your choice of varieties to your circumstances.

A greenhouse certainly helps in the propagation and subsequent cultivation of young plants in the spring, but it is by no means essential. On the contrary, these may easily be raised by anyone who has room to spare in a window indoors; then by June they should be good plants that may either be grown on indoors or used in various ways outside during the summer.

Many varieties, for instance, are ideal for hanging baskets both outside and in the greenhouse, and to add to these there are many others that will be equally effective in window boxes and other plant containers. Others can be used in the normal way as summer bedding plants, or as tall 'dot plants' to relieve the monotonous appearance of lower-growing summer flowers; and although there are comparatively few fuchsias that can be planted permanently outside there are plenty more that will make superb temporary specimens for planting out in summer until they are lifted and stored for the winter.

The training of the plants to different shapes also adds enormously to their interest. Apart from the uses just mentioned they may be grown as 'standards', which are in effect miniature trees, with a bushy head on top of a length of bare stem (grown in this way they are ideal for tubs or large pots as well as for using as specimens in the open garden). Or they may be grown in a similar way but on a smaller scale, to be used as striking temporary table decorations. Other forms include the pyramid, pillar and espalier, of which details will be found in Chapter 8 on Training to Shape, together with some of the most useful varieties to choose for these purposes.

From all this it will be seen that the fuchsia is an extremely accommodating plant; it will even stand a certain amount of neglect and faulty treatment. But it should be pointed out that it well repays careful cultivation and attention to detail, particularly in the training of the different forms just mentioned. There is nothing like the production of healthy, well-shaped and free-flowering plants for bringing pleasure and satisfaction to the grower and the little extra work involved in producing these is really neither here nor there.

It is however no use trying to run before you can walk and the beginner will certainly do well to start off with the easier varieties and with the simpler forms of training. Success with these will then encourage him or her to go on to the finer and more capricious varieties and the more elaborate forms of training; and finally perhaps to the exhibition bench, although it must be remembered that here the standard is extraordinarily high—you have only to go to one of the major fuchsia shows to see what can be really achieved.

Finally, it will pay, right from the start, to join one of the Fuchsia Societies that exist in Britain. Apart from being a fruitful source of information and advice this will foster enthusiasm and also help to keep you up-to-date with the modern varieties. These are constantly arriving, in an endless stream, and while some are here today and gone tomorrow some stay the course for many years. Some of those mentioned in this book have indeed been cultivated for well over half a century, which indicates their utter and complete reliability. It is therefore well worth considering some of these as possible plants to start with, before going on to the more difficult ones.

2
The Fuchsia Family

Before going into the characteristics of the fuchsia in more detail it may be as well to clarify what is meant by the terms 'genus', 'species' and 'variety'. In so much of the literature of gardening it seems to be taken for granted that everyone, even the absolute beginner, knows what these mean, but in my experience they actually convey very little to the gardener who is not botanically minded.

To speak of the family of fuchsias, as in the title of this chapter, is not strictly correct, as in botanical nomenclature the term 'family' covers a number of different groups of plants often differing widely in appearance but bound together by one or more common characteristics; thus, for a simple example, all the different groups in the family *Compositae* have flower-heads that are more or less daisy-like; they include such very different plants as the common daisy of our lawns and the large-flowered chrysanthemums that brighten our gardens and greenhouses in autumn. The fuchsia, as it happens, belongs to a much less well-known family, *Onagraceae* (formerly *Oenotheraceae*), which also includes such apparently different plants as the clarkia and godetia.

Each of the groups belonging to a family is called a genus (plural genera). A genus comprises all those plants which in addition to possessing the common features of the family also share certain others which are not found in the other groups in the family. The first name of any plant is always the name of the genus and it remains constant throughout the whole group, much in the same way as our surnames serve to identify a particular group of people. In all members of the genus there is usually a distinct family resemblance, so that in the genus Fuchsia, for instance, it is possible to recognise the different kinds without much trouble, even though they may vary considerably in some ways.

Each genus is in turn divided into 'species'. These represent the different natural forms of the plant and each one includes all those individuals that are so much alike that they can be given the same

name. This name, the 'specific epithet', is always the second name of
the plant and it may be likened to our Christian names, although in
the case of plants it is applied to a group of individuals and not merely
to one. In some cases this specific epithet is the Latinised form of the
name of the person who first discovered the plant, but more often it is
derived from the place of origin—as in *Fuchsia boliviana*, from
Bolivia—or from some particular feature of the plant, as in *F.
procumbens*, meaning prostrate. The name of the genus followed by the
specific epithet forms the 'specific name' of the plant.

A hybrid is the result of crossing two species in the same genus. To
identify such hybrids the specific epithet given to them is preceded by
the multiplication sign ' × ' and followed by the specific epithets of the
two parent plants connected by an ' × ' and enclosed in brackets e.g.
F × bacillaris (microphylla × thymifolia). True hybrids of this type are
however rare among fuchsias, in which most of the plants in
cultivation are 'varieties' or, as they are now correctly called,
'cultivars'. In botanical nomenclature the term 'variety' is also used in
other ways; but for gardeners, and in this book, it refers to those
plants bred in cultivation, by hybridisation or some other means,
which remain the same throughout subsequent propagation and which
are sufficiently different from others of their kind to warrant a separate
name. This name, which is never in Latin form, is always a vernacular
or 'fancy' one, and it should be enclosed in single quotes—as in 'Rose
of Castile' and 'Tennessee Waltz'—although in catalogues and older
books these quotes are often omitted.

The Fuchsia in Nature

The genus Fuchsia covers about a hundred known species, which apart
from a few found in New Zealand and Tahiti, are indigenous only to
Central and South America, from Mexico southwards, including some
of the West Indies. It may therefore seem that they are practically
tropical in their requirements but in fact their needs are governed more
by their immediate habitat than by their geographical locale.

Some species, for instance, are found in the forests, where although
the air is warm the plants are protected from the full heat of the sun by
the canopy of trees and the humid atmosphere. Others grow more in
the open, but generally in damp, shady situations at quite high
altitudes where they are cool but safe from frost; and one species, *F.
magellanica*, grows in the wet, windy conditions of the extreme south,
so is quite hardy in other temperate areas.

Wherever they grow, however, the species have one thing in common: they are all hardwooded, so that unlike herbaceous plants they retain their stems and branches from year to year. In most cases this means that they develop into typical shrub-like plants, but a few form trees, sometimes as much as 30ft high, while others develop into no more than trailers or creepers. Of these last some are epiphytes, which means that they grow on trees without being actual parasites; instead they obtain their moisture from the rain and humid atmosphere and their food from the fallen leaves, bird-droppings and other debris that collect among the branches.

Many of the *species* are still known only in their wild state. Indeed, only about a quarter of them are in cultivation and even these have been largely ousted by the innumerable *varieties* that have been produced for more than a century. Like their ancestors these varieties are more or less shrubs, but when they are cultivated under glass little use is made of this natural habit; instead the plants are normally cut back annually, either to keep them within bounds or, more commonly, to provide the cuttings that within a year will replace the parent plant with others that are more vigorous and free-flowering.

The Flower

The fuchsia flower (Fig 1) consists basically of a calyx tube terminating at the outer end in four distinct segments (sepals); four petals, separated from each other but together forming the corolla; and the sexual organs comprising the stamens and pistil.

In describing the colours of the bloom it is usual to refer to the tube and sepals as 't' and 's', and to the corolla as 'c', but while these abbreviations may be clear enough to the knowledgeable grower they are not quite so easily understood by the beginner. This is largely due to the fact that on the fuchsia the sepals and the petals forming the corolla are quite different in appearance from those of most common flowers, on which only the petals are decorative and colourful while the sepals forming the calyx are merely green and of little ornamental value.

In the fuchsia, on the other hand, the sepals, which serve to protect the petals and sexual organs during the bud stage, form an important and colourful feature of the bloom, as apart from often being quite large and colourful they may be of almost any colour other than blue. At the bud stage these sepals are in fact the only visible colourful feature of the flower, but as they expand they become either spreading

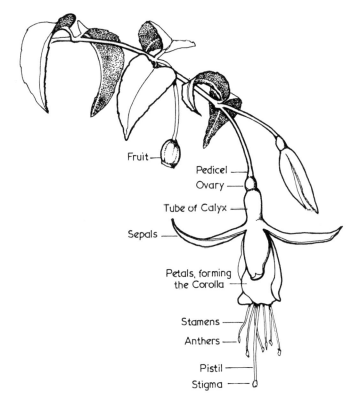

Fruit

Pedicel

Ovary

Tube of Calyx

Sepals

Petals, forming
the Corolla

Stamens

Anthers

Pistil

Stigma

1   Structure of the fuchsia flower

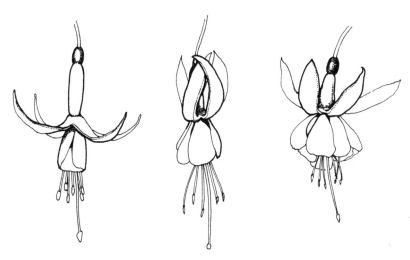

2   Different types of fuchsia bloom: left, single with spreading sepals; centre, single
with reflexed sepals; right, double with recurving sepals

or reflexed (Fig 2), and thus reveal the corolla of petals, in which an enormous range of colour is found. It may indeed be of any shade or colour other than yellow, which so far has eluded the skill of the hybridiser.

It should be mentioned here, incidentally, that in catalogues and other sources of information the descriptions of fuchsia flowers do not always agree about the colour. This may be partly due to the fact that different people see the same colour in different terms, but a further explanation is that fuchsia flowers of the same variety can vary considerably in colour according to the soil, temperature and situation. Thus plants grown in the shade tend to be paler in colour than those grown in full sun, while those grown outside are often of a richer and deeper shade than those grown under glass.

Fuchsia flowers may also be single, double (Figs 1 and 2) or semi-double, but the doubling always refers to the petals, never the sepals. The term 'doubling' is in any case somewhat misleading as the number of petals in a flower other than a single can vary considerably; sometimes a semi-double can be scarcely distinguished from a single, while a full double may have so many petals that the corolla bears little resemblance to a normal one. Things are not made any clearer either by the fact that some of the additional petals may not develop fully. To make matters worse there are also 'petaloid' flowers in which some of the stamens have become converted to immature petals (an unexplained change which when carried to its full extent accounts also for the doubling of many other flowers).

Each flower is carried on a slender and usually drooping stalk ('pedicel'), terminating in an ovary (Fig 1), which may be red or green, at the base of the tube. In this ovary are the 'ovules' which will eventually become seeds after pollination and fertilisation, processes which are facilitated by the special arrangement of the stamens and pistil. In most fuchsia flowers these organs project beyond the corolla but the pistil to a greater extent than the stamens carrying the pollen on their anthers (Fig 1). Any movement of the pendent flowers, perhaps caused by the wind or by insects, tends to shake the pollen from the stamens and as it falls it alights on the sticky surface of the stigma at the tip of the pistil below it. In this way pollination and eventually fertilisation are effected.

The flowers of the fuchsia are usually but not always produced singly in the axils of the leaves (where the leaves join the stem), but in some species they are produced in clusters. Of the tender fuchsias commonly grown, the best example of this latter type is to be found in

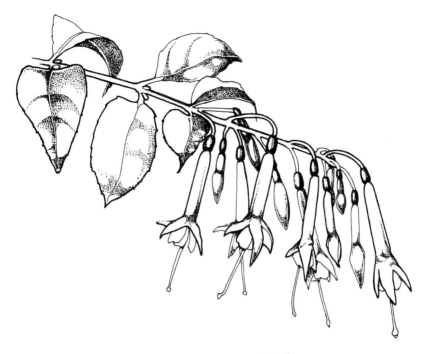

3　Flower cluster of a *triphylla* hybrid

a group known as the *triphylla* hybrids (Fig 3) derived largely from *F. triphylla*, but although these are sometimes grown and make excellent plants they have nothing like the colour range of the other and more popular varieties.

With their varied shapes and colourings fuchsias are undoubtedly among the most beautiful of our cultivated flowers but they lack one attraction—fragrance. This is not even found on any of the many species, perhaps because owing to the structure of the flower there is no need to attract insects to pollinate the plants. The chances of a fragrant fuchsia being produced are therefore fairly remote, but—who knows?—one may eventually turn up and become the source of a further virtue in these beautiful and fascinating plants.

The Leaf

The general shape of the fuchsia leaf is more or less oval but it is not the same on all the different species and varieties. On some the leaves are much narrower or more rounded than on others, and they may have either a pronounced terminal point or none at all. Both toothed and plain-edged ('entire') leaves are also found, and while the colour is predominantly green, purplish or reddish variations are also found.

15

These variations may be found to some extent on many of the different cultivated varieties, but while the leaves on these tend to be of roughly the same size they vary widely on the different species from less than an inch to several inches long.

There is one unusual feature about the fuchsia leaf and that is the possession of certain pores known as 'water stomata'. On all plants the leaves in a sense 'breathe' through somewhat similar pores, but on the fuchsia the purpose of these water stomata is to provide an outlet for any excess moisture that the plant has taken up through its roots. Thus when a plant is fully turgid after being well watered, and the weather is so humid that the normal pores are not capable of coping with all the excess moisture, these water stomata come into operation, with the result that minute drops of moisture can be seen on the edges of the leaves, where they are often mistaken for condensation. There are, incidentally, many other plants with the same or similar arrangements for disposing of excess moisture, among them being poppies, the arum lily, and such well-known house plants as the spider plant (*Chlorophytum elatum variegatum*), the rubber plant (*Ficus elastica*) and busy lizzie (*Impatiens sultanii*).

The Fruit

After pollination and fertilisation, the flower falls, leaving behind the fruit, which on the fuchsia is a fairly large berry (Fig 1) divided internally into four compartments, each containing a number of seeds. These berries, usually red, green or brown in colour, form readily on plants which have not had their faded flowers removed, but there is little point in allowing them to develop except where hybridisation is to be carried out, as the production of seed only weakens the plant.

There is however one species, *F. procumbens* (see page 33) on which the bright red berries, each about three-quarters of an inch long, make a valuable addition to the decorative effect of the plant; and fortunately this trailing or creeping species is one of the few that are hardy enough to risk outside as permanent garden plants in most parts of Britain. It is one of those suitable for use in rockeries.

The fruits of several species are both pleasantly edible and rich in vitamins, and in their natural home these are gathered as a delicacy by the native Indians. It is worth quoting from an old gardening book, *The Cottage Gardener's Dictionary*, published in 1859, which states quite confidently:

16

When gardeners discover the way to improve the size and flavour of fruits, we cannot doubt but that those of the Fuchsia and Cactus will be among the first novelties in the dessert.

Perhaps the writer had in mind the large purple fruits of *F. corymbiflora* or the green ones of *F. fulgens*, for both undoubtedly make excellent eating. But despite this he has so far been wrong, and it is unlikely that we shall ever see the fruits of the fuchsia being sold as a commercial proposition, unless of course other foods begin to run short!

3

Fuchsia Hedges

In fuchsia catalogues and books you will find a number of varieties listed as hardy, but these would be better described as 'near-hardy'. The test of a truly hardy plant is that it must come unharmed through the worst winters in the coldest parts of the country, and very few fuchsias can be relied upon to do this. There are, however, plenty that will survive the worst winters that we are likely to get in Britain with only their top-growth—the parts above ground—damaged, so that after the winter they soon send up new shoots from the ground. This is the type of fuchsia that books and catalogues mean when they speak of 'hardy fuchsias'.

The degree of damage suffered in winter depends in any case on several factors. In a mild winter there may be little or no harm done even in cold areas; and in the milder parts of the British Isles, where even the worst winters are not so severe as those in less favoured areas, the plants may never come to any harm at all. The actual variety of fuchsia makes a difference too. An increasing number of varieties is proving hardier than was once believed; research into this aspect of fuchsia growing is continually going on, and there is no reason why the amateur should not carry out his own experiments with any surplus plants—the fuchsia is so easily propagated that several of these are usually available. They must, however, be planted in early summer rather than in autumn so that they become thoroughly acclimatised before the winter. Even then it is not advisable to judge by one season alone. It is a well-known fact that plants become hardier as they age, so that a fuchsia which manages to survive one or two mild winters will probably come safely through a much harder one later on.

Another factor which affects the reliability of the fuchsia as a garden plant is the supply of moisture. No fuchsia will live for long without water, and although there is usually more than enough of this in a British winter the soil may become so frozen that the plant cannot take it up (a possible reason why fuchsias die in winter even though they may be hardy enough to stand considerable frost). But at least you can guard against this lack of moisture due to frost, by making sure that the soil around the plant is sufficiently protected to prevent it freezing hard—a point which will be dealt with later in this chapter.

Fuchsia Hedges

Anyone who has seen the beautiful fuchsia hedges of south-west England, the west of Scotland or parts of Eire can hardly be blamed for fancying a similar one in his or her own garden, but here the choice of variety is a vital one—obviously it is no use planting a hedge that is going to be cut down by frost—even occasionally, in a hard winter.

In fact only two species make up most of the hedges in the milder parts of Britain and these are definitely the ones to use for a reliable hedge in all parts of the country. One is *F.Riccartonii*, the hardiest of all our cultivated fuchsias, named after the Riccarton Gardens in Edinburgh, and the other is *F.magellanica*, named after its native home in the area bordering the Straits of Magellan at the wet and windy tip of South America, where the weather can be at least as bad as any British winter. Both of these can be relied upon to come through most winters unharmed in most parts of the country, and if they do happen to be cut down by exceptionally severe weather they will soon recover and make strong hedges again in the following summer, when *F.Riccartonii* will grow as much as 5 or 6ft.

On both these species the flowers are of much the same colour, with the tube and sepals scarlet and the corolla violet, although on *Riccartonii* the corolla is a slightly deeper shade and the sepals are noticeably broader. Apart from this the main difference lies in the size and vigour of the plant, with *Riccartonii* being the larger and stronger-growing.

F.magellanica has also given rise to several varieties, all hardy enough to grow as shrubs or even as hedges in the milder areas. *F.m.alba*, with shorter white flowers tinged with mauve, is one of these, and another is *F.m.gracilis*, much the same as *magellanica* itself but of more slender habit. Then in addition there are two fine variegated forms, one being *variegata*, with the leaves margined with creamy-yellow, flushed pink, and the other *versicolor*, a variety with slender stems bearing striking grey-green leaves, rose-tinted when young and variegated with creamy-white as they age; both have flowers the same colour as the type plant. These variegated varieties make excellent foliage shrubs, but unfortunately neither of them is quite as hardy as *F.magellanica* itself.

19

'Near-Hardies' as Garden Shrubs

The fact that fuchsias of one sort or another are commonly seen growing as shrubs in gardens shows that some are hardy to all intents and purposes, although as they are not completely reliable everywhere they are actually 'near-hardies'. These are the ones that are usually listed as 'hardy fuchsias' in books and catalogues—the main ones among them are listed later on in this chapter—and there is a much wider assortment of these than of the varieties suitable for hedging.

Whether they are actually hardy or not, though, they make invaluable flowering shrubs for the garden. From the latter half of the summer until well into autumn, as the other flowering shrubs gradually get scarcer, the fuchsia really comes into its own. It thrives in the cool misty atmosphere of autumn and will go on producing a splendid show of glowing colour until the frosts arrive; if the weather is exceptionally kind it may even go on flowering right up to Christmas.

The most obvious place for garden fuchsias is in the shrub border, where they will not only keep the show of colour going until well into the autumn but also get some winter protection from their neighbours. They can be equally well used in the herbaceous border, where if they do happen to be cut down in the winter they will be no worse off than the ordinary herbaceous plants. Some will do well too on top of a retaining wall, or by the sides of paths or gateways, while others can be used to brighten up the surface of a blank wall.

There is in fact virtually no place where the fuchsia will not feel at home. Even a shady spot will suit it well enough, and there are often angles in the house wall that will provide ideal shady situations for it. And for screening-off compost heaps and other unsightly features there are few shrubs that can beat the fuchsia, with its long period of flowering.

Preparing the Site

The actual preparation of the ground, planting and subsequent cultivation of outdoor fuchsias is much the same, whether they are being used as shrubs or hedges, the first thing being to prepare the site thoroughly. Most shrubs and hedges, of course, do best in a position fully open to the sun, but this is of little importance with the fuchsia, which does not object to partial shade; in fact a sheltered site in the shade may be better than a more exposed one in full sun, particularly where cold winds are likely to be a source of trouble.

Of more importance is the need for perfect drainage, so that the plants do not stand too wet and cold through the winter. On the other hand, as has already been pointed out, the fuchsia must never go short of water. To guard against this dig the ground deeply and thoroughly, preferably to a depth of about 18in, but take care not to bring the subsoil to the surface. For hedges this digging should extend to at least 18in on either side of the hedge, while for shrubs a circular site about a yard across is about right.

At the same time as the digging is being done, plenty of leafmould, garden compost or other humus should be worked in as deeply as possible on light soils, while on heavier ones grit or coarse sand can be used to keep the soil more 'open'. Plenty of humus is needed near the surface too, to provide immediately the cool moist root-run that fuchsias enjoy. It will help the initial root-action if a good sprinkling of bonemeal is incorporated at the same time.

On chalky or alkaline soils—those containing a high proportion of lime—good dressings of humus are even more important, as they will help to create the more acid conditions that fuchsias like. On such soils the best method of all is to replace the soil with a more acid one from elsewhere, but this will probably not be practicable and the use of plenty of humus provides the best alternative.

Ideally all this preparatory work should be done in the autumn so that the soil has time to settle during the winter. However, if it has to be done in the spring adequate firmness can be obtained by treading the soil prior to planting, whenever it is dry enough not to clog.

Planting

With the soil ready, the next step is the actual planting which will have to be done when the plants are available from the nurseryman, usually in late May or early June. If they arrive earlier then they will need to be thoroughly hardened off before being put in, except perhaps in the case of really hardy kinds which will probably have been grown outside at the nursery.

For hedges, the spacing of the plants will depend on several factors, including the vigour of the species or variety used, the required density of the hedge and whether or not quick results are wanted. To produce a dense hedge quickly, a spacing of 18in apart should be about right, but if more time can be allowed and fairly vigorous kinds are used, a distance of 3ft apart should not be too much to provide what is in effect a close row of bushes rather than a very dense hedge. Incidentally, if cost is a factor to be considered, you can economise, if you do not mind waiting a year or two, by buying only two or three plants to start with. These will yield plenty of cuttings the first year and as the fuchsia is so readily propagated with these (see Chapter 10) you can soon have more than enough to make a good hedge.

All the fuchsias intended for permanent outdoor planting are normally grown in pots at the nursery, and before they go in they must be thoroughly watered and allowed to drain. Those in clay or plastic pots can then be easily extracted: first place one hand over the top of the pot, with the stem of the plant between the fingers; then invert the pot and give its rim a gentle tap on the edge of a bench, or on the handle of a spade inserted in the ground, so that the plant falls safely into your hand with its roots intact in the soil. If the plants are in paper or other temporary pots it is usually safe to plant them just as they are, although the roots will emerge more easily if the pot is first torn down in two or three places.

The plants are then ready to go in and at this stage the main thing is to plant them sufficiently deeply. The deeply dug, humus-rich soil will help the roots to find their way deep down out of reach of frost and drought but they can be further helped by deep planting; the top of the 'soil-ball' from the pot should be 4 or 5in below the surface. As well as assisting the roots this will protect the lower growth buds during subsequent winters. Firm planting is essential, but the roots must not be suffocated and the best method of avoiding this is to pour a little fine dry soil round them first before finally filling the planting hole and treading the soil down gently but firmly.

After-care

After planting, very little further attention is needed during the first summer, apart from the usual weeding of the plants and removal of the dead flowers. In very dry weather watering will also be needed and this is a job that must be done thoroughly. As soon as the soil shows signs of drying it should be soaked to a good depth, with another soaking following as it starts to dry again. However, in long spells of dry weather watering can be reduced by mulching the ground with compost, leafmould or bracken after the first soaking, so that the soil is protected from the drying sun and wind.

For the first year the plants will have adequate food in the soil and thus no feeding will be needed; in fact feeding in this early stage may do more harm than good by discouraging the roots from growing and fending for themselves. For older plants, an occasional feed with any of the general fertilisers on the market is advisable, but rich manures and nitrogenous fertilisers should not be used: they will result in soft growth that will not stand the winter.

Summer Pruning

Flowering hedges of any kind cannot of course be kept as severely trimmed as the non-flowering privet, hawthorn, beech and so on, or many of the flowers would be lost. All that is needed in the first summer is to trim back into shape any long, wandering shoots; then in May of the following year any dead wood should be cut out and the hedge given a fairly hard trim-over with the shears. This will leave plenty of time for the new shoots to grow and flower at the usual time in late summer.

Fuchsias grown as shrubs are dealt with in much the same way. They will usually take on their own natural shape, but badly placed growths may be removed and any sparse branches can be stopped to thicken them up.

Winter Care

With each successive year the roots of the plants will go deeper and thus be more out of harm's way, while the plants themselves will become hardier as they age. For the first year at least, though, the new plants will need a little protection, which can be supplied by covering the roots and the base of the stems with ashes, leafmould, bracken or soil. This covering should be at least 4in thick over a distance of about a foot all round the plant; and if plenty of bracken can be worked in among the base of the stems to provide extra protection so much the better. These coverings should be left on right through the winter until severe frosts have finished, probably about late April or early May.

No pruning should be carried out in autumn or winter. Left as they are, the old stems provide still further protection until there is no further risk of severe frost. Then any dead ones can be cut out at ground level, and any side-shoots on the surviving ones can be shortened back to two leaf-joints from the main stem. By doing this new growth will be encouraged, with consequently a better display of bloom in the summer.

In cold areas, or after a particularly hard winter, some of the plants may look dead, but there is usually no need to worry. Unless the frost has gone deep enough to kill the roots new growth is sure to start up again from the ground in early summer, and the plants will then flower just as well as those that suffered no damage.

So far as the fuchsia hedge is concerned, one of its drawbacks is that in winter it lacks leaf or flower, and looks pretty stark. For brighter effect a few plants of the winter jasmine (*Jasminum nudiflorum*) can be incorporated in it. This plant is not a true climber but with a little help it will scramble up through the fuchsia branches and provide a mass of evergreen shoots liberally studded with its bright yellow starry flowers right through the winter. It is an easy and very hardy grower and a good stock can soon be obtained either by layering the long shoots—these often layer themselves—or by putting some of the flowering shoots in water in winter, when they will soon root in a warm room.

Varieties to Use as Shrubs

Apart from the choice of colour the main thing to consider is the ultimate height of the plants. This is of course most important where they are being grown in a border, as it is no use having large ones at the front and smaller ones at the back. But it is well worth considering wherever the plants are to grow. Large plants near a pathway, for instance, can be a nuisance, particularly in wet or wintry weather, while if they are to be grown under a window they must not be tall enough to interfere with the light.

The following list includes most of the 'near-hardies' that are commonly grown and easily obtained. Others are sometimes listed as hardy, but while it is always worth taking a chance on these it is always safest to rely on the well-tried ones.

'Achievement'   A very old variety going back nearly a century and one that is usually grown indoors or under glass. It is however sometimes listed as hardy, and its rapid upright growth, combined with its early flowering, make it well worth a trial. The tube and the recurving sepals of the single flowers are cerise and the corolla purple.

'Amy Lye'   Another very old single that is also sometimes listed as hardy. It makes a strong upright plant that can also be grown as a standard or pyramid, but in the garden it makes a medium-sized free-

The popular, all-purpose 'Princess Dollar', suitable for indoors, summer bedding and for growing as a standard. Its flowers are cerise and purple.

Old fuchsias that have been stored for the winter should be cut back to two or three buds in early spring (*Author*)

Before repotting, the old soil is carefully removed (*Author*)

Cuttings are taken when the young shoots are two or three inches long (*Author*)

Stopping young fuchsias: left, pinching out the tip of the stem; centre, a stopped plant; right, a plant left unstopped for growing on as a standard (*Author*)

'Madame Cornellison', with
scarlet and white flowers, is
suitable for hedges in mild areas

'Mrs Popple', with scarlet and
violet-purple flowers, is hardy
enough to make a fine shrub
even in the North

flowering plant with foliage of a medium green. The rather stubby tube and sepals are a waxy-white and the corolla rosy-cerise.

'David'   A very hardy variety of short upright growth that makes it ideal for the edge of a border. Also ideal for window boxes. The small single flowers, cerise on the tube and sepals and purple on the corolla, are freely produced on a compact plant that needs no stopping.

'Edale'   A single that makes a free-flowering upright bush. The sepals are pink and the corolla violet.

'Florence Turner'   Ideal for a sheltered shady position. A very hardy variety of excellent colouring, with the china-rose tube shading to white at the tip of the sepals and the corolla magnolia-purple.

'Hollywood Park'   An American variety that is usually grown under glass but which some catalogues list as hardy. Makes an upright bushy plant with a profusion of semi-double flowers, cerise on the tube and sepals and white, tinged with pink, on the corolla.

'Howlett's Hardy'   A very popular 'near-hardy' stocked by most fuchsia nurseries. One of the larger-flowered of the outdoor fuchsias, with a profuse show of the single blooms, scarlet on the tube and sepals and purple on the corolla. Makes a medium-sized bush.

'Lena'   A very old variety going back to Victorian times. One of the most versatile of all fuchsias, as apart from being hardy enough in most areas it can be trained to most shapes under glass. The semi-double flowers appear in profusion on a rather arching bush some 2ft high and are quite large for such a hardy variety. The tube and sepals are pink and the corolla orchid-purple.

'Madame Cornelissen'   An old variety that makes a good hedge in mild areas. It has proved to be one of the best 'near-hardies' for both colour and size of the bloom. Makes a strong upright bush with long buds that open to single flowers, with a white corolla and scarlet tube and sepals. Does best when given some protection each year.

'Mrs Popple'   One of the hardiest of the 'hardies', capable of withstanding even northern conditions. Makes an upright or rather spreading bush with plenty of medium-sized flowers, scarlet on the tube and sepals and deep violet on the corolla.

'Nicola Jane' A comparatively recent introduction among the 'hardies' and one that has proved unusually reliable. The double flowers are light scarlet on the tube and sepals and white, veined pink, on the corolla. Growth is strong and upright.

'Phyllis' A variety which has proved hardy in all but the coldest areas. Will grow to 4 or 5ft even when cut down in a hard winter and makes a profuse display of semi-double blooms. Almost a rose-coloured self, with the corolla a deeper shade. A rapid grower, it can be used as a hedge in mild areas.

'Port Arthur' Over a century old, this variety has proved quite hardy in mild areas, although it is more commonly used as a bedding plant. Makes a strong bushy plant with plenty of medium-sized blooms on which the tube and sepals are red and the corolla bluish-purple.

'Prosperity' A fairly hardy fuchsia of good colour, with the tube and sepals of the double flowers a waxy-scarlet and the corolla rose-red. Not commonly listed.

Other varieties worth trying in the garden will be found in the general list of varieties at the end of the book.

Fuchsias in the Rock-Garden

No doubt the alpine 'purist' will turn up his nose at the thought of fuchsias being included among his precious treasures, but nevertheless for the average gardener they can provide some bright colour in the rock-garden just when it is most needed, from late summer onwards. Obviously the larger kinds are out of the question in the average small garden's rockery but there are several suitably small species and varieties that, planted in little groups, will provide a splendid show.

The great risk, as with the larger ones, is winter damage, but in the well-drained conditions of the rock-garden they should be reasonably safe; and if they are planted where they can get their roots well under a stone so much the better, as this will help to protect them from both the frost in winter and the sun in summer. It will also help if those fuchsias that are planted permanently are intermingled with some of the creeping rock-plants such as thyme, aubretia, stonecrops and rock-phloxes as these too will provide a little protection.

Planting

There are two ways of treating fuchsias in the rock-garden. One is to use them as permanent plants, setting them straight into the soil; the other is to plant them in their pots, so that they can be lifted and stored for the winter.

The former method obviously makes less work, and with the hardier varieties it is usually safe enough in all but the coldest areas; and even there a few cuttings taken in summer will provide a safeguard against total loss. Planting in pots, however, has distinct advantages. If the plants are potted into 5in pots before they go in, they will have plenty of fresh soil to last them right through the summer, by the end of which they should have made considerable growth. The range of possible varieties is increased too, as there is no need to restrict yourself to the hardy ones—the tenderer greenhouse and indoor ones are available too. The cascade and trailing types can, for instance, be very effective grown over a large stone or over a dry wall, where long after the usual rock-plants are over they will make a splendid show.

Planting is best done, as with other hardies and near-hardies, about the end of May or early June, when those that are to go straight into the soil should be put in quite deeply, with the base of the stem well covered. This will help to keep them safe from frost but, as in other areas of the garden, to be on the safe side a thick top-dressing of soil, peat, leafmould or ashes should be placed over the roots towards the end of the year. Apart from this very little attention will be needed, apart from watering in dry weather and removing the faded blooms.

When planting in pots the main thing is to give the plants a good watering before they go in—if they are planted when very dry any future watering runs outside the pot and leaves the plants high and dry. As the plants will eventually be lifted, deep planting is not important, but to create a natural effect the rim of the pot should be below ground. No winter top-dressing will be needed, of course, and the only thing to watch is that the plants are lifted and stored well before there is any risk of severe frost.

Varieties to Use in the Rock-Garden

'Alice Hoffman' Forms a dwarf, compact and upright bush with bronzy leaves and a plentiful supply of small single flowers, cerise on the tube and sepals and white on the corolla. A useful variety that can be grown almost anywhere.

31

F.bacillaris   A Mexican species that is more of a curiosity than any-thing else. It forms a compact, low shrub, with a dense growth of branches carrying small, almost stemless, leaves and tiny rose-red flowers no more than ¼in long. It can be obtained only from fuchsia specialists and is too tender for permanent planting.

'Brilliant'   A very old variety with a long flowering season. It is too tender for permanent planting, but planted in pots it will make a very bushy upright plant with plenty of single flowers, the tube and sepals scarlet and the corolla violet-magenta.

'Corallina'   A rather large plant for the rock-garden, but a good one for a dry wall. It forms a low, arching bush with unusually large flowers for a hardy variety and with attractive greenish-bronze foliage. The tube and sepals of the single flowers are bright scarlet and the corolla rich purple.

'Display'   See general list at the end of the book.

'Elsa'   An early-flowering variety that is hardy in mild areas. Makes a strong, rather sprawling plant with light green foliage and a profuse show of medium to large flowers, ivory-pink on the tube and sepals and rosy-purple on the corolla. Well worth planting out in pots for an early show.

'Flash'   See general list at the end of the book.

'Lady Thumb'   A sport from the well-known 'Tom Thumb', with which it is identical except in colouring. The tube and sepals are bright red and the corolla white.

'Magellanica discolor'   A dwarf variety of the hardy *F.magellanica*, the same except in size.

'Peter Pan'   An excellent hardy dwarf shrub, with a profuse display of small red and purple flowers.

'Princess Dollar'   Also known as 'Dollar Princess'. The small double flowers, cerise on the tube and sepals and rich purple on the corolla, are produced very freely on this small upright bush. Not hardy but well worth planting out in pots.

F.procumbens An interesting New Zealand species with small upward-pointing flowers consisting of an orange-yellow tube and sharply reflexed greenish-purple sepals. There is no corolla but the red and blue stamens form an attractive feature. This is a creeping plant, rooting as it goes, and its small plum-like fruits add an additional touch of colour in the autumn. It is interesting to note that although the flowers of this species come the nearest of any to being a pure yellow it has so far proved impossible to use it as a source of yellow in other varieties. The plant is hardy only in very mild areas.

'Prostrata' A low spreading plant with single flowers, on which the tube and sepals are red and the corolla violet. Ideal for the rock-garden where the small but plentiful flowers will make a fine mass of colour. Hardy in all but the coldest areas.

'Pumila' A very old variety going back to early last century. Forms a miniature bush smothered in small single flowers, the tube and sepals scarlet and the corolla mauve. An excellent rock-garden variety in all but the coldest areas.

'Reflexa' Hardy enough only for the mildest areas, this makes a delightful little shrub with tiny leaves and small cerise flowers, followed by black fruits. Possibly a form of *F.bacillaris*.

'Thornley's Hardy' A very prolific variety with single flowers, waxy-white on the tube and sepals and cerise on the corolla. Makes a good dwarf plant in most areas.

'Tom Thumb' A hardy dwarf that is well over a hundred years old but still popular. It makes a short upright plant with plenty of medium-sized flowers, light carmine on the tube and sepals and mauve-purple on the corolla. It flowers late, sometimes not until August, and although it may suffer damage in winter it usually manages to come through.

4

Fuchsias for Bedding

In Victorian times, and up to the First World War, the fuchsia was extensively used as a bedding plant in the gardens of great estates and country houses. But those days are gone, and fuchsias are not all that often seen as bedding plants except in public parks and gardens. One reason for this is the initial cost of the plants, if you have to buy the comparatively large number required for bedding, and another reason is that people just do not think of the fuchsia as a bedding plant nowadays.

A reasonably large bed of fuchsias is in any case impractical in the average modern garden, for apart from the cost there is just not room for it. But even a small bed can be very effective. A small circular one in the lawn or in a patio can be striking, particularly if an ordinary or 'weeping' standard is placed at its centre (the cultivation of 'standards' is dealt with in Chapter 8); or a small bed of fuchsias at the side of a path or beneath a window will provide something a bit out of the ordinary. The initial cost of the comparatively small number of plants required for such a bed is hardly likely to be prohibitive; if it is, you can start off with one or two plants and soon work up the necessary number from cuttings, which will make good plants in the same year that they are taken (see Chapter 10).

Preparation of the Bed
A bed that will grow most ordinary bedding plants well should be equally satisfactory for fuchsias, if prepared in the usual way. However, if the bed is a new one it should be well dug before treading it to a fine firm tilth, at the same time working in plenty of compost or old manure plus a sprinkling of general fertiliser. The best effect will be obtained if it is raised in the centre so that a perfectly flat appearance is avoided. This is easily done by merely raking the soil towards the middle, while at the same time taking care to create a regular slope on every side.

Planting

This is done about the end of May or early June, when there is no further risk of frost. Young plants in 3 or 3½in pots are readily available about then, but whether you use these or some of your own raising the planting is exactly the same. Make sure that they are moist at the root first, then plant them firmly, with 15in from plant to plant. There is no need to bury them deeply, as is done for the hardy fuchsias, as they will be lifted and stored for the winter. As long as the 'soil-ball' from the pot is well covered they will be deep enough for the summer months.

Very little attention is needed after planting, but if the young plants are single-stemmed, with no sign of branching, they should be 'stopped' by pinching out the tip of the stem as soon as growth is well away. Watering too will be needed in dry weather but after the initial watering-in to settle the soil around the roots the plants should be allowed to dry a little to encourage new root-action. Provided the roots are moist a light spray overhead in the evenings of warm days will in fact do more good than watering the soil.

Varieties for Bedding

A bed of one variety is far more effective than a mixed one, but it must be the right type of variety. A good bedding fuchsia must have a long flowering period; it must be reasonably weatherproof; and it must have a compact, upright habit. Some suitable varieties with all these qualities are listed below, together with a few fuchsias with variegated foliage. These are invaluable for providing a really decorative edge to the bed if you like a contrasting border to it; they should be pinched out fairly frequently to ensure a good show of the colourful leaves.

'Army Nurse' A fine fuchsia for all purposes other than baskets. Makes a good bedding plant and in mild parts of the country is quite hardy. Forms a strong upright bush with a profuse display of medium-sized semi-double flowers, deep carmine on the tube and sepals and with the corolla violet-blue, pink-tinged. Under glass it is useful for training as a standard or pyramid.

'Brilliant' Well over a hundred years old, this variety is still well worth growing. There are plenty of medium-sized flowers, scarlet on the tube and sepals and violet-magenta on the corolla, and they make a very early show on plants used for summer bedding. Under glass it makes an excellent standard.

'Charming'   A compact and bushy fuchsia with foliage of a bright yellowish-green. The tube and sepals of the single flowers are carmine and the corolla reddish-purple. Extremely weatherproof. Can also be grown as a standard under glass.

'Dollar Princess'   Also known as 'Princess Dollar'. A popular plant for market and one that is invaluable for its early flowering and ease of cultivation. A good 'house-plant' for growing indoors. The tube and sepals of the medium-sized double flowers are cerise and the corolla a rich purple. Can also be grown as a standard or pyramid.

'Elsa'   Hardy in mild areas this variety makes an excellent summer bedding plant. There are plenty of medium to large semi-double flowers and the foliage is light green. The tube and sepals are pink and the corolla rosy-purple. Early flowering. Makes a fine standard in the greenhouse.

'Forget-Me-Not'   A very old variety with small single flowers which make up in number what they lack in size. The tube and sepals are a pale flesh-pink and the corolla is pale blue, with the plant itself making a strong upright bush.

'Kolding Perie'   A Danish variety of unknown origin, this makes a fine bedding plant with an abundance of medium-sized flowers, the long tube and sepals waxy white and the corolla pink with shades of salmon. Makes a vigorous upright bush.

'Liebriez'   A continental variety with small double flowers. The tube and sepals are pale cerise and the corolla a pinkish-white with pink veinings. Has all the good points of a bedding fuchsia – long flowering period, etc.

'Marin Glow', a strong indoor favourite which can also be grown as a standard or pyramid *(Harry Smith)*

Fuchsias can be grown in an endless variety of containers for effect—here 'Leonora' is placed on an elegant table to decorate a patio *(Michael Warren)*

'Madame van der Strasse' The tube and sepals of the double flowers are a pale cerise and the corolla is flushed and veined pink on this well-tried bedding variety. Makes an upright bushy plant with plenty of medium-sized blooms. Very weather-resistant.

'Moonlight Sonata' A good single-flowered variety, with the tube and sepals bright pink and the corolla light purple flushed with pink at the base. Makes a strong upright and free-flowering bush with an abundance of medium-sized blooms. Remains in bloom for a long period.

'Nicola Jane' An attractive hardy variety, suitable for bedding, with double flowers that are completely weatherproof. A strong upright plant, with the tube and sepals scarlet and the corolla white, veined pink.

'Port Arthur' A very old variety that is sometimes listed as hardy although it is not reliable everywhere. An excellent summer bedder as it makes a very sturdy bush with a liberal display of medium-sized blooms. The tube and sepals of the double flowers are red and the corolla bluish-purple.

'Tolling Bell' A strong upright and compact bush which produces its single flowers very freely. The tube and sepals are scarlet and the bell-shaped corolla white. An excellent bedder that also makes a good standard.

Foliage Varieties for Bedding
'Burning Bush' Forms a low bush with the foliage attractively variegated golden-bronze and red. The flowers are of medium size, with the tube and sepals scarlet and the corolla purple.

'Cloth of Gold' Primarily a foliage variety, with the few flowers much the same in colour as 'Burning Bush'. Makes a bushy plant with golden foliage, red on the underside. Made all the more attractive by its cascade habit.

'Lena Dalton', a good upright, has a pink tube and recurving sepals blending with a pink-flushed blue corolla *(Harry Smith)*

'Golden Treasure'   A very old foliage variety going back more than a hundred years. It produces its scarlet and purple-magenta flowers very freely but it is grown mainly for the beautiful red-veined golden foliage. Of strong, very spreading habit it has been used to fill whole beds in the past, with the golden foliage making a complete mat with the red veining showing up prominently.

'Green-n-Gold'   As its name implies the foliage on this variety is variegated green and gold. The single flowers, produced in clusters, are a coral-pink self and are liberally produced on an easy plant of vigorous spreading habit. Flowers over a long period.

'Sunray'   Another old variety going back more than a hundred years but still in demand. The single flowers are small, salmon-pink on the tube and sepals and rosy-purple on the corolla. But it is the foliage that is the main feature of this plant. Variegated with pale green, cerise and creamy white, all with a silvery cast, this makes an unusually effective display on a plant that makes a good upright bush.

The Triphylla Hybrids for Bedding

F.triphylla, from the West Indies, has been extensively used in the hybridisation of several varieties that are invaluable for bedding; in fact one of them, the orange-scarlet 'Thalia', has been widely used for this purpose for over a century.

These hybrids are quite different in appearance from the usual type of fuchsia. The flowers each consist mainly of a long tube, with only a diminutive corolla, and they are produced in clusters at the ends of the shoots, while the foliage is a distinctive green, strongly tinged with olive or bronze tints.

Suitable for both the greenhouse and bedding, the different varieties all form compact upright plants. Unfortunately they have one failing: they are among the most tender of fuchsias, so that they cannot stand any frost at all. Care must therefore be taken to lift those used for bedding well before there is any risk of frost. For this reason they are better bedded out in their pots rather than being actually planted in the ground, as they can then be lifted at a moment's notice.

A piece of slate or asbestos under each pot will prevent the roots from escaping from the pot into the soil and this will induce the plants to flower all the more freely; then when they are lifted they can go into a heated greenhouse to flower until probably about Christmas. Then, like other fuchsias, they should be gradually dried and stored.

'Billy Green' Makes a strong erect bush with profuse clusters of salmon flowers against olive-green foliage. A fairly recent addition to the *triphylla* hybrids.

'Coralle' An orange-salmon self. The terminal clusters of flowers show up well against the blue-green foliage and the plant makes a medium-sized bush of erect habit.

'Gartenmeister Bonstedt' An old and well-tried variety with orange-scarlet blooms in profusion. Makes a vigorous upright plant with dark green foliage shaded with purple.

'Heinrich Heinkel' A light rose-pink flower that shows up well against the dark olive-green leaves, deep magenta on the underside. The growth tends to be more spreading than that of most *triphyllas*.

'Leverkusen' A not very popular hybrid with cerise flowers. It makes a strong upright and bushy grower with plenty of bloom and is thus a good variety for bedding.

'Mantilla' The flowers are a deep carmine on this *triphylla* hybrid of cascading habit. Although useful for bedding it is equally good for growing in a basket in which it will make an unusual show with its tubular flowers and bronzed foliage.

'Thalia' Perhaps the best of the bedding *triphyllas*, forming a strong upright bush with orange-scarlet flowers against olive-green foliage. With its vigorous habit it soon makes a good bush.

'Traudchen Bonstedt' An excellent greenhouse plant but equally suitable for bedding. The light sage-green leaves, veined with a lighter shade, make a fine setting for the salmon-pink flowers, produced on an upright bushy plant. One of the best of the *triphyllas*.

'Trumpeter' A more recent introduction with geranium-lake flowers darkening with age. The growth is upright and bushy and the foliage is a rich bluish-green.

5

Fuchsias for Growing in Containers

So far I have dealt only with fuchsias for the garden but many city-dwellers and some of the residents of smaller towns have no garden to speak of, and this need not prevent them from growing fuchsias. Everyone has some sort of space: it may be only a backyard or its more resplendent relative, the patio, or it may be just a balcony or porch. Whatever or wherever it is, it will enable the owner to enjoy the beauty of these lovely flowers. In fact, where there is no garden all the more attention can be paid to the fuchsias and all the various methods of growing and training them as described in Chapter 8.

If you have a garden, of course, so much the better. You can still grow fuchsias in several other ways than in the soil of the garden. In baskets, tubs, troughs or window boxes they will add tremendously to the appearance of your home as well as giving you endless interest.

Hanging Baskets

Probably the most usual way of growing fuchsias, apart from in the garden soil, is in hanging baskets. There is virtually nowhere that a place for these cannot be found, and once in bloom they make a spectacular show. If you do not want to make the baskets up yourself most nurseries will do it for you, but if you buy the young plants or have some of your own it is quite a simple job to tackle. Young plants in 3 or 3½in pots are what are needed, and if it has not already been done these should be 'stopped' at three or four pairs of leaves by pinching out the growing point.

The baskets can be made up in May, although if there are no facilities for keeping them safe from frost the job is best done about the end of the month, when they can go straight outside. The number of plants required depends on the size of the basket, with four or five, according to their vigour, going into a 15in one and correspondingly fewer in the smaller sizes, down to say a single plant in an 8in basket.

Fuchsias are often mixed with other plants in hanging baskets but for the best results they should be used on their own, preferably in one variety only.

Making-up the Basket

To simplify this operation the basket should be placed in the top of a bucket so that it cannot rock about. It should then be lined either with a thick layer of sphagnum moss (obtainable from nurserymen or florists) or with stout polythene sheeting, which can be camouflaged with a thin layer of moss beneath it. These linings are absolutely necessary with the wire baskets of the traditional type, but in the more modern ones made of rigid plastic the only preparation needed is to place a few broken pieces of pot over the drainage holes to keep them clear.

If polythene sheeting is used as the lining, a few holes should be made an inch or two up from the bottom, so that while most of the excess water can drain away there will still be a little left in the bottom to act as a reservoir; if moss only is used, a similar effect can be obtained by placing a saucer in the bottom of the basket on the inside of the moss lining. Apart from reducing the amount of drip these 'reservoirs' will save a certain amount of watering, which is one of the main tasks with baskets, particularly when they are hung outside. Exposed to the sun and wind they dry out very quickly, so that constant watering is needed, possibly twice a day in hot weather. If a basket should become so dry that the plants are flagging the only thing to do is to take it down and soak it from underneath in a basin of water.

The next thing is to place a layer of compost in the basket. Either John Innes No 2 or one of the soilless composts will do for this. The plants are then knocked out of their pots and placed in the basket with their roots towards the centre and the top-growth pointing outwards and upwards. Further compost is then well worked in round the roots with the fingers, so that no empty spaces are left to form air-pockets; and finally the surface should finish with a saucer-like depression to facilitate watering. The job of watering can also be made easier by putting a rim of clay, disguised with moss, around the edge of the basket.

After the basket has been well watered and allowed to drain, it is ready for hanging, but first make sure that both the hook and the support to which you attach it are strong enough: a basket of plants, particularly when it has just been watered, is a considerable weight. In a small and fairly frail greenhouse or porch, for instance, no more than one or two quite small baskets should be suspended from the roof unless an additional support in the way of a vertical post from the floor to the ridge is provided; if a large basket is to be grown, say for exhibition purposes, it is better stood in the top of a chimney-pot or oil drum rather than hung from the roof.

Subsequent Treatment

The first 'stopping' at three or four pairs of leaves should ensure with many varieties that the plants branch out sufficiently and the only thing necessary is to make sure that these branches are seen to advantage. Varieties with a stiff, arching habit can be safely left to look after themselves, but the branches of the 'cascade' types tend to hang almost straight down on top of each other, and with these some branches should be left to grow naturally and the remainder should be separated from them by a 'fence' of string supported by short canes inserted in the soil, pointing outwards and upwards. Apart from supporting the flowers this will also make the basket look bigger.

Watering has already been mentioned but feeding will also be needed as the plants grow. Any good proprietary fertiliser, preferably in liquid form, should give satisfactory results, as long as the rule of 'little and often' is observed instead of relying on occasional heavy doses.

A careful watch will also have to be kept for greenfly and other pests, but these are dealt with in Chapter 11.

A Selection of Basket Varieties

'Bouffant' A very free-flowering cascade variety with large, long flowers on which the corolla opens wide. The tube and sepals are red and the corolla white, heavily tinged with rose. Single flowers.

'Caballero' Forms a lax bush with plenty of large double flowers. The tube and sepals are a deep salmon-pink and the corolla, which is surrounded by petaloids, varies from bluish-purple to violet, with splashes of salmon-pink. The petaloids surrounding the corolla vary from white to red. A striking basket variety.

'Cascade' A very popular variety which soon makes a large basket. The fairly large flowers are borne in profusion on a plant of 'cascade' habit. The tube and long narrow petals of the single flowers are white and carmine and the corolla is deep carmine.

'Crackerjack' A lax bush with a liberal display of large, long-tubed single flowers. The tube and sepals are white, faintly flushed pink and the long petals of the corolla are pale blue.

'Fire Opal' A semi-double making a willowy trailer, with the tube and sepals red and the long corolla dark blue. A good variety for a basket in the shade.

'Fluffy Frills' A recent introduction. The double pink flowers are a medium-pink self, with the corolla ruffled and frilly. Semi-trailer.

'Golden Marinka' A natural cascade variety although it can be trained to almost any shape. The semi-double rich red flowers are of medium size and prolific enough to show up well against the colourful foliage of light green splashed and barred with gold.

'Indian Maid' The richly coloured double flowers are produced very freely on a willowy plant that makes a good trailer. The tube and sepals are scarlet red, the latter being long and recurved, and the long corolla is a royal purple aging to wine-red.

'Jack Shahan' Makes a strong-growing bush varying from a spreading bush to a cascade habit. The large single flowers are produced very freely and are practically a self-coloured rose-bengal.

'La Campanella' The tube and sepals are white and the corolla imperial purple on this bushy trailer. The semi-double flowers are of medium size and are produced very freely.

'Lassie' This is more or less a trailing form of the popular 'Swingtime', with much the same colouring, a bright red tube and sepals and white corolla. The lax habit makes it suitable for a basket.

'Lilibet' A fine basket variety with long pointed buds that open to long-tubed flowers with recurving sepals. The tube and sepals are white, flushed pale carmine, and the corolla is soft rose with deeper shading. The double flowers are very freely produced.

'Marinka' An extremely popular basket variety with a lax habit that can be trained to almost any shape. The single flowers are almost a red self and are of medium size, produced in great numbers.

'Mrs Marshall' Over a hundred years old this variety is still among the most popular. It naturally forms a strong upright bush but it can be easily trained to a cascade or any other shape, and the single flowers, waxy-white on the tube and sepals and rosy-cerise on the corolla, are produced very freely.

'Niobe' Both the colouring and the very fast-growing habit of this splendid variety make it ideal for basket work. Of cascade growth it soon fills a basket with double flowers, white flushed pink on the tube and twisting sepals and an intense smoky-rose shade on the corolla.

'Red Jacket' With large double flowers in profusion this strong-growing cascade variety makes a beautiful basket plant. The tube and sepals are bright red and the large fluffy corolla is made all the more attractive by the long projecting dark red stamens.

'Red Spider' Another vigorous cascade fuchsia with a profuse display of long large single flowers. The tube and recurving sepals are a deep crimson and the corolla is deep rose-madder margined with deep crimson. A very showy fuchsia.

'Stanley Cash' A fairly recent introduction with large double flowers on a plant of trailing habit. The tube and sepals are white and the corolla deep royal-purple.

The above are only a few of the countless number available and many more will be found in The Varieties, page 139.

Window Boxes

These provide another popular way of growing fuchsias, whether or not you also have them in a garden. Most houses and flats have one or more windows that can be used for them, and filled with fuchsias they can make a glorious show of colour right through the summer.

To avoid losing any light the box may if necessary be screwed to iron brackets fixed to the wall beneath the window but whether it is on these or on the actual windowsill the main thing is to make sure that it is safe, particularly where upstairs windows are being used. In very high and exposed positions it may be advisable to secure the box to the window-frame with hooks and screw-eyes, which will both keep it secure and enable it to be easily removed for maintenance when it is not being used. Not that even a gale is likely to dislodge a heavy box of soil sitting firmly on its base, but there is always a chance that it may be accidentally moved nearer the edge where it could eventually become dangerous.

The length of the box will obviously depend upon the size of the window, but in width and depth it should not be less than 9in. Ready-made boxes in various materials are of course available but any handy-man can soon knock one up, preferably using a wood such as oak, teak, redwood or cedar, finished with the appropriate oil. The wood for the sides and ends should not be less than ½in thick, while that for the bottom should be about an inch, and all the timber should preferably be dressed with a wood preservative harmless to plants, such as green Cuprinol. As the windowsill will probably have a slight outward slope a thin batten of wood will be needed under the front of the box to bring it up level. A double row of ½in holes will also be needed in the bottom of the box to provide drainage.

Planting the Box

The drainage holes in the bottom must first be covered with a layer of 'crocks' (pieces of broken clay pots) placed concave side downwards over them, so that water can get through but soil cannot. This is then covered with a 2in layer of moist peat, or better still with pieces of well-rotted turf placed grass-side downwards, and the box is then ready for filling with compost; this job is best done when the box is in position, as a filled box is very heavy to move.

The compost can be either John Innes No 2 or a soilless one. Or if you have some good soil in the garden a suitable one can be made of three parts soil and one each of peat and coarse sand or grit, plus a sprinkling of general fertiliser. It should be placed in the box to about an inch from the top, taking care to firm it well down, particularly at the edges and corners.

As with baskets, planting is best done about the end of May or as soon as there is no risk of frost. As purchased plants may have come straight from a heated greenhouse they will have to be hardened off: keep them indoors for a few days near an open window, which should be closed at night at first and then gradually left open for a time.

The actual planting is done in the same way as for baskets, with young plants from 3 or 3½in pots being used. A box of the width mentioned will take a double 'staggered' row, with say a row of a trailing variety being put in at 9in apart along the front of the box and some dwarf upright ones at the back of the spaces between them. If they have not already been 'stopped' this should be done by pinching out the tip of the stem at the third or fourth pair of leaves.

See that the plants are well watered before putting them in and that the upright ones are fastened to short canes to prevent them from being blown about. Later on it may be necessary to change these canes for rather longer ones if the window box is in a very exposed position and the plants need securing.

Although window boxes are in the open it is no use relying on the rain to water them; in fact in the shelter of the wall they will receive very little, except when it is blowing straight on them, and even this will not wet the soil very much. A constant watch must therefore be kept on the soil so that as soon as it shows signs of drying it can be given a good soaking.

An occasional feed with a proprietary fertiliser will be needed as soon as the box is well filled with roots but this should not be overdone or the plants may produce more leaf than flower. A fertiliser containing a good proportion of potash is the most suitable to use as this will not only encourage free flowering but also make the plants more weather-resistant.

Varieties for Window Boxes
Any of the trailing and cascade fuchsias mentioned in the previous section on hanging baskets may be used at the front of window boxes, and the following list includes only some of the smaller upright ones for the back of the box. Further varieties will also be found in the list of varieties at the end of this book.

'**Athela**' A free-flowering single of upright bushy habit. The tube and sepals are creamy-pink and the corolla salmon edged with pale pink. Also makes a good plant for indoors and for summer bedding.

'**Beacon**' Forms a compact self-branching bush with single flowers. The tube and sepals are red and the corolla is mauve-pink, with the petals opening out flat.

'**Bridesmaid**' A double with the tube and sepals white flushed pale carmine, and the corolla pale lilac-pink, deeper coloured at the edge of the petals. The large blooms are freely produced and the habit is compact and upright.

'**Carnival**' A low, bushy variety with double flowers, waxy-white on the tube and sepals and a dull, flat red that belies the name on the corolla. The flowers are large and plentiful.

'Citation'   A popular single, with large flowers in profusion. The tube and sepals are rose-pink, with the latter curling well back, and the corolla consists of four broad white petals, veined with pink at the base, which open out to a bell shape. Forms an upright well-branched bush and may also be grown as a standard.

'Esther'   A good fuchsia for a window box in the shade. Makes a medium-sized bush with double flowers with a white tube and frosted white, green-tipped sepals. The corolla is purple-violet.

'Icecap'   A sturdy medium bush with bell-shaped single flowers that last a long time. The medium-sized blooms are cardinal red on the tube and sepals and the corolla is white veined red. The stocky habit makes it a good pot-plant.

'Iced Champagne'   An old variety of short-jointed, self-branching habit. The single flowers, dawn-pink on the tube and sepals and rhodomanthine-pink on the corolla, are produced quite freely. Makes a good pot-plant.

'Iris Amer'   A fine variety with medium-sized flowers in profusion on a short-jointed upright bush. The tube and sepals of the double blooms are white flushed pinkish-orange and the corolla is bright red with splashes of carmine and orange.

'Mr A. Hugget'   A variety of unknown origin that does well under most conditions. An upright grower with small single flowers, cerise on the tube and sepals and mauve-pink on the corolla.

'Mr W. Rundle'   A popular market variety with medium-sized flowers on upright branches. The tube and sepals of the single flowers are pale rose and the corolla orange-vermilion. An old variety that does well in a window box and indoors and also makes a fine standard in the greenhouse.

'Petite'   An easy grower with a very long season of bloom. The tube and sepals are pale rose and the corolla lilac-blue fading to lavender. The small double flowers are prolific on an upright bushy plant.

'Pink Bon Accord' A fine easy variety for the window box. Although the flowers tend to be small they are very profuse on a neat, short-jointed upright plant. The single flowers are a pale pink on the tube and sepals, the corolla deep pink.

'Plenty' A recent introduction forming a close, short-jointed bush with plenty of medium-sized single flowers. The tube and sepals are two shades of pink and the corolla is violet-purple.

'Powder Blue' A good variety for the house and windowsill, particularly where the latter is in partial shade. The colouring of the medium-sized blooms is very attractive, with the tube and sepals rose and the corolla pale blue.

'RAF' Forms a bushy plant with a good display of double flowers. The tube and sepals are red and the corolla powder pink. Very showy when in bloom and makes a fine half-standard.

Other Containers

Apart from hanging baskets and window boxes there are plenty of other containers that can be used very effectively in the garden and on the patio or balcony, and any of the foregoing fuchsia varieties plus many others can be used in them.

Wooden containers are among the most versatile, and are widely used in America. They can be bought or built in almost any shape or size, and are long lasting if made out of durable woods such as redwood, cedar, or cypress. Fir, pine, or plywood can also be used if treated with wood preservative. Most weather to an attractive grey.

Wood containers prevent overheating of the soil, and are naturally porous. Air and water pass through the sides, cooling and oxygenating the roots. The boxes or tubs must have adequate drainage to prevent overwatering. They should be set on blocks to allow for air circulation, or built with short lifts or legs. One version is the popular 'plant trough', which is more or less a window box standing on legs. This will take several plants of both trailing and upright fuchsias, the pots in which they are growing being merely stood in the 'trough'. To keep the soil in the pots evenly moist and to reduce the amount of watering they should be packed round with moist peat or moss, which will also improve the general appearance.

The trailing fuchsias make ideal plants for any hanging type of receptacle, made in various materials to be suspended from the roof or ceiling. The only snag with these is that the plants must be removed for watering, but this does not present much of a problem as long as they are allowed to drain before they are returned to their container.

Outside, almost anything that will hold sufficient soil and has sufficient drainage holes can be used for fuchsia plants. Old chimney pots, plain or painted, are ideal for trailing fuchsias, although with these it is best to grow the plants in a basket placed in the top of the pot. The same method can be used too for old oil-drums, which can soon be made more attractive with a coat of paint. And for fastening to the wall there are the old iron hay-racks, if you can get hold of one, which provide a fine way of growing quite a number of plants. If they are lined first with 1 inch mesh wire-netting and then with moss, they will hold the necessary soil quite satisfactorily and once planted up they will make a splendid show.

Garden urns and vases are perhaps a more obvious choice and they will make excellent receptacles for both upright and trailing fuchsias, which can be planted straight into the soil in them as there is usually adequate drainage. Painted wooden wheelbarrows and various types of wicker basket can also be used but in these it is best to grow the plants in pots to avoid using so much compost. There is plenty of scope for ingenuity in the use of containers, and on the balcony and patio particularly there is no end to the variety of utensils that can be used to give an unusual and decorative effect.

6

Fuchsias Indoors

So far I have dealt only with the growing of fuchsias out-of-doors, but probably most people, in both town and country, would like to have some indoors as well. The fuchsia certainly makes a splendid house-plant. But, as the centrally heated and well-lit houses of today do not provide ideal conditions for it, it is not one of the easiest; indeed it was much more at home in the darker, damper and cooler conditions of the Victorian home, in which it was one of the most popular indoor plants in summer.

Many of the fuchsias grown indoors in Victorian times are in fact still with us and are still being sold in shops and markets. The more modern varieties may be far showier but they lack the constitution of these old ones, and for sheer simplicity and reliability it is the old ones that you should try to get. Several of them are mentioned in the list below and in general they are fairly easy to obtain.

When the plants are purchased they may be in either 5in or 3½in pots. Those in the larger size may be left in the same pots for the rest of the summer—they usually come on the market about May. Those in the small pots, however, will need potting on into the larger size as soon as the roots can be seen in the drainage holes. This potting-on is dealt with in detail in Chapter 7.

Light, Temperature and Humidity

The fuchsia does not like a strong light indoors. A sunny window is no place for it and it will be far safer in a north-facing one or in a part of the room out of direct sunlight. Not much heat is needed either, but fortunately when the fuchsia is in season central heating is not being used to any great extent. A more likely cause of trouble then is draughts from open windows, so make sure the plant is draught-free.

The greatest enemy of fuchsias indoors is too dry an atmosphere. It thrives best in a cool humid one, so wherever possible it should be stood on a moist base such as a tray of moist peat or wet gravel to provide the necessary humid micro-climate. An occasional mist spray, particularly on dry sunny days, will also help to keep the plant healthy, but if, despite all precaution, the leaves and flowers start to fall the best thing is to put the plant outside for a time, where it will appreciate the cool, fresh air and the more humid nights. But make sure it is put where it cannot blow over.

Watering and Feeding

As with all house-plants, watering is the most difficult part of looking after indoor fuchsias. They do not like a waterlogged soil, but they do like a moist one, although they will stand a certain amount of dryness without coming to any harm. There are various ways of telling when a plant is too dry: the soil may look and feel quite dry; the pot may feel light in comparison with that of a wet plant; and if the leaves look limp, the plant is in sore need of water. But there is no need to rely on these indications—it is better to stand the plant in a saucer of water, which you refill when it is not only empty but quite dry. Then the plant will take up just as much water as it needs and no more.

If anything it is better to err on the side of under, rather than over, watering. As long as the plant is not in the sun, which may cause the leaves to flag even though the soil is quite moist, it will come to no harm if the leaves do become a trifle limp, provided that you do not leave it for too long. Just feel the leaves every day and at the first sign of limpness give the plant a thorough soaking from underneath.

Greenfly are likely to be a nuisance but these can easily be kept at bay by using one of the aerosol insecticides at the first sign of them. Never let them get a good hold, as apart from spoiling the plant they may also spoil any polished surface on which the plant may be standing. This is due to the sticky 'honeydew' which the insects excrete and which drips down to the surface beneath. Greenfly and other pests are, however, dealt with in more detail in Chapter 11.

Winter Care

Although fuchsias will sometimes go on flowering until well into the winter this will only weaken them for the following year, and it is best to bring them to a halt before then by gradually reducing the watering until all the leaves and flowers have fallen. During this period the plant should be stood in a sunny spot outside, not only to avoid having the fallen leaves and flowers all over the place but also to expose the plant to the sun and air so that the branches and shoots will ripen or harden into firm brown wood ready for the winter. The treatment then should be as described in Chapter 9.

Varieties for Indoors

'Ballet Girl' An old double variety going back to Victorian times and still popular. The crimson-cerise tube and reflexing sepals and the full corolla, white veined red, amply justify the name and the plant is well worth trying as both bush and standard.

'Berliner Kind' Another Victorian one. Forms a low-growing bush with very double flowers of medium size, with the tube and sepals scarlet and the corolla pure white.

'Bon Accord' Tube and sepals creamy-white, corolla pale purple, flushed white. An unusual variety in that the profuse single flowers stand erect. An easy grower that makes a bushy plant of very upright habit.

'Grus aus Bodethal' An old but showy variety with single flowers freely produced. The tube and sepals are a bright crimson and the corolla a very deep purple that lightens with age. Not an easy variety to obtain these days.

'La Honda' A fairly modern fuchsia that makes a good exhibition as well as indoor plant. It is easy to grow and forms a strong upright bush with dark green glossy foliage, against which the semi-double flowers, red on the tube and sepals and rosy-purple on the corolla, show up well.

'Leonora' A very easy fuchsia with plenty of medium-sized bell-shaped flowers on a bushy, free-branching plant. A soft pink self.

'Marin Glow' A fuchsia of strong upright habit that apart from being a good indoor one can also be grown as a standard or pyramid. It is easy to grow and very free-flowering, with single blooms waxy-white on the tube and sepals and imperial purple on the corolla.

The simple beauty of 'Mission Bells' *(Harry Smith)*

'Mission Bells'   A fuchsia of simple beauty, with the tube and sepals a bright red and the wide-open corolla having attractively wavy-edged petals of rich purple. The medium to large single flowers appear very early and make a splendid show on a strong upright plant.

'Pink Delight'   A small and compact grower with double flowers that are almost a pink self. There are plenty of medium-sized flowers on a free-branching plant.

'Pink Jade'   Makes a neat short-jointed bush with a liberal show of medium-sized flowers that last well indoors. The tube and sepals are pink, shading to green at the tips, and the saucer-shaped corolla is orchid-pink with a picotee edge.

'Rose of Castile'   A very old variety of strong upright growth that is sometimes listed as hardy although it is more often grown indoors or under glass. Needs plenty of food and water and given these it makes a fine free-flowering plant with a profuse show of small single flowers against light green foliage. The tube and sepals are white and the corolla purple, flushed with pink.

'Rufus'   A turkey-red self that grows quickly into a strong upright bush with plenty of medium-sized flowers over a long season. Also grown for exhibition and makes a good standard.

'Scarcity'   A very old variety that has been popular as a house-plant since Victorian times. Contrary to its name, it produces a wealth of single flowers, rather stout and short in shape. The tube and sepals are cerise and the corolla rosy-purple. It is said to be hardy in many areas.

'Southgate'   An invaluable variety with all the qualities of a good fuchsia. Easy to grow, it makes a vigorous upright plant with masses of double flowers, almost a pink self in colour.

'Swanley Gem'   A fine fuchsia that has been popular for over eighty years. A popular plant for exhibition but also makes a good house-plant. Forms an upright bush with the tube and sepals of the single flowers a rich scarlet; the corolla, which opens out flat, is violet.

'Flying Cloud' shows the pleasing, bushy shape which makes it a good standard. The flower is ivory white except for a tinge of pink on the tube *(Michael Warren)*

'Temptation' An excellent fuchsia of strong upright growth with long, crinkly leaves. The medium-sized flowers are borne in great profusion and the tube and long sepals of the single blooms are white, the corolla orange-rose.

'The Doctor' An easy grower forming a good bush. There are plenty of medium-sized flowers, with the tube and sepals flesh-pink and the corolla rosy-salmon.

7

Greenhouse Cultivation

Compared with the number of varieties mentioned for the garden and indoors, the number for the greenhouse is infinite. They come in a constant stream, both in Britain and from America, so that nowadays there are many hundreds available; there is no attempt to list them in this chapter as all of those in the List of Varieties at the end of this book can be used for greenhouse work. Some of them can be trained in several different ways, as standards, pillars, pyramids, espaliers and climbers, and these are duly indicated in Chapter 8, 'Training to Shape'. But they can all be grown in the most popular form of all—the pot-grown bush—and in this chapter only this form is discussed.

Some varieties are easier to grow than others and the beginner will do well to start with these. Many fuchsia nurseries offer 'beginners' collections' and one of these can provide a good starting point. Then after some experience with these the grower will no doubt be tempted to go on to the more difficult ones and to the more complicated methods of training.

Making a Start

Fuchsias are usually available as either rooted cuttings, supplied more or less all the year round except in mid-winter, or as young plants in small pots or soil-blocks from spring until the end of June. The latter obviously produce flowering plants earlier in the season, but for the economically-minded the much cheaper rooted cuttings, if obtained in early spring, will still give good results—only a little later. In either case the plants will be in bloom by the summer and they can then be used to provide further cuttings during the summer and in the following spring.

The time when the plants should be obtained in spring depends upon whether the greenhouse is heated or not. When ordering them—and this should be done well in advance, preferably in late autumn—this information should be given to the nurseryman so that he will be able to send them at the right time. For a cold house this will be about mid-March or a fortnight later in cold areas, but if the greenhouse can be kept at a temperature of at least 45°F (7°C) a start may be made as early as mid-February.

Treatment of Young Plants in Pots

If young plants in pots or soil blocks are obtained, these will be practically ready for moving on into larger pots as soon as they arrive; but it is advisable to give them a chance to recover from the packing and travelling first. If they are at all dry they should be stood out on the greenhouse bench with enough space between them to allow a good circulation of air, and given a good watering. If a moist base of peat, sand or gravel can be provided so much the better, as this will help to create the humidity that they enjoy. Protect them from full sun by suspending a piece of butter muslin or clear plastic sheeting over them.

As the plants will be well rooted in the pots there is little risk of over-watering. The pots should be filled to the rim each time the compost shows signs of drying on the surface, while those in soil-blocks should be lightly watered overhead. To reduce the amount of root-watering needed the plants should be regularly syringed overhead and around the pots.

If for any reason a plant should flag from want of water it will come to no harm as long as it is thoroughly soaked again, but once it has reached this stage it is no use relying on one watering only. The pot should be filled to the rim at least two or three times to make sure the water soaks right through to the bottom; or, better still, the pots should be stood in a tray of water until the moisture has seeped through to the surface.

After a few days the plants will have fully recovered from their travels and be ready for the final potting.

Treatment of Rooted Cuttings

These need more care than the young pot-grown plants. The main thing is to get them potted-up as soon as possible when they arrive, and for this a supply of pots and compost should be put ready beforehand. Three-inch pots, either clay or plastic, will be amply large enough, and there will be no need to provide drainage in these in the form of 'crocks' or broken pieces of pot. A suitable compost for these pots is John Innes No 1, or alternatively one of the proprietary soilless composts.

Half-fill a pot with compost, then pull it to one side while the rooted cutting is held in the centre of the pot, its roots dangling over the compost. Add further compost to fill the pot almost to the rim. One or two light taps on the bench and light finger-pressure of the compost against the roots will then make it adequately firm and at the same time leave sufficient space at the top for watering. On no account should the compost be compressed too much, particularly if a soilless one is being used; the aim should be to make it just firm enough to hold the cutting securely.

These rooted cuttings are then treated in much the same way as the young pot-grown plants dealt with above. As soon as they are potted they should be given a thorough watering-in and then allowed to become fairly dry before being soaked again. (Extra care is needed here if a soilless compost is being used, as if this becomes too dry it will be very difficult to make it thoroughly wet again.) By allowing the plants to dry a little after potting, root action is encouraged; and then after a further soaking they can be watered normally.

Bush Plants

Most of the fuchsias grown in greenhouses are bush plants—plants with a single stem breaking into several branches. Although some fuchsia varieties will branch out naturally without any special treatment most of them need to be 'stopped' or 'pinched out'. Young plants bought in pots will probably be already stopped, but if not both these and the rooted cuttings should be dealt with as soon as they have made three or four pairs of leaves.

This stopping consists merely of removing the tip of the stem (Fig 4) with the finger and thumb back to the topmost pair of leaves. This will result in side-shoots springing from the axils of the leaves (where each leaf joins the stem). The strongest of these side-shoots should again be stopped when they have made two pairs of leaves, while the weaker ones can be left unstopped so that the growth is evened out. No further stopping is needed as any more would delay flowering, which takes place about six weeks after the final stopping.

Note, however, that this method of stopping applies only to plants being grown in the bush form. Other forms, such as the standard, pillar, pyramid and so on, require different treatment as described in Chapter 8.

4   A cutting stopped at three pairs of leaves

5   Young bush plant with the two longest shoots stopped

Final Potting

A few weeks after being potted the rooted cuttings will be pushing their roots through the drainage holes in the pot and by this time they will be at a stage roughly corresponding to that of the young plants in pots despatched by nurserymen. From then on the treatment of both sorts is the same, the main thing being to get them into their final, flowering pots before they become root-bound in the small ones.

As the plants sent out by nurserymen are often in pots less than 3in, it is sometimes recommended that they should first be moved into 3 or 3½in ones, but although this saves space for a time it makes unnecessary work and also increases the risk of the plants becoming too dry (plants in small pots can be very difficult to keep sufficiently moist). A better method is to pot them straight into the 5in pots in which they will flower, for with a little care this can be done without any risk and with definite benefit to the plants. This 5in size is usually adequate for the first season, although after that larger ones may be needed.

Either clay or plastic pots may be used for this final potting, but nowadays the latter are usual. Good drainage is essential in either kind. For plastic pots a layer of the rough material left after sieving peat and soil will be sufficient, but for 'clays' use pieces of broken pot, with one large piece placed concave side over the drainage hole and covered with smaller pieces.

The compost may be either John Innes No 2 or one of the soilless ones, but this is not to say that more ordinary mixtures cannot be used. In Victorian times, long before these more sophisticated composts were introduced, superb plants were grown in the only materials available, which were mainly soil, leafmould, old manure and soot and wood ashes; and even an up-to-date American publication recommends a compost consisting of no more than equal parts of coarse sand or sandy soil, leafmould and cow manure. Fuchsias are in fact not fussy about the soil they are grown in as long as it is well-drained, spongy and rich. It should also be free from lime, as the fuchsia does not like an alkaline soil.

The evening before the potting is to be done the plants should be given a good watering so that the roots and soil (the 'soil-ball') are quite moist when they go into the new compost; otherwise if the 'soil-ball' is dry any water that is applied after this potting will merely bypass it and thus delay new root action, if no worse.

In doing the actual potting the plant should first be knocked out of its pot as described in Chapter 3 and then placed on enough compost in the new pot to bring the top of the soil-ball to about an inch below the rim. New compost is then added until the soil-ball is buried about ½in deep, when one or two taps on the bench and a moderate firming of the compost round the soil-ball will settle the plant in at the right depth to allow for watering from the top.

After the potting the plants must be thoroughly watered and left to become almost dry before being soaked again. This will encourage new root action and after a further watering followed by a similar spell of dryness it will be safe to start normal watering by filling the pot to the rim whenever the soil surface looks dry. Overhead spraying and a humid atmosphere will also help to produce strong growth, but if this and the watering happen to be neglected for any reason, with the result that the plants flag a little, there is no need to worry. A thorough soaking will soon perk them up again and they will come to far less harm than if they are over-watered at this early stage.

Subsequent Cultivation

As soon as the plants are obviously making new growth they may need staking, although this is not usually necessary on those of a fairly stiff, upright habit. Those of a more lax or spreading growth will however certainly need it, as otherwise the eventual weight of bloom may drag the branches down and even break them off. The simplest and least unsightly way of staking such plants is to loop all the stems together with soft green twine tied to a central split green cane (obtainable at garden shops) but in doing this care should be taken to preserve the natural grace of the plant, as it is only too easy to bundle all the stems together so that neither flowers nor leaves show off to advantage.

As growth progresses, more and more water will be needed until by the time the plants are well in bloom it will be almost impossible to over-water them; the easiest and safest way then is to stand each one in a pot-saucer (these are obtainable in the same diameters as the pots) which should be refilled as soon as it is empty. Normally this will be about once a day, but at the height of the summer even this may not be sufficient and a second daily dose may be needed.

The need for so much watering can be reduced to some extent by providing both shade and high humidity. There are various ways of shading greenhouses, but for people who are free to attend to their plants all day there is no better method than some form of blind, particularly when there are only spasmodic spells of sunshine. Those made of narrow wooden slats and fitted to the outside of the greenhouse are perhaps the ideal type, but they are expensive, and for a cheaper sort there is a spring-loaded roller-blind made of green plastic sheeting for fitting to the inside of the greenhouse. Cheaper still are the various types of screening, generally made of some plastic material, that can just be thrown over the greenhouse roof; the disadvantage of these is that they are apt to blow away if they are not securely weighted down or fastened in some way.

For people who are away all day, a more permanent form of shading will be necessary, and for this a thin paste of flour and water, applied to the outside of the glass, is as good as anything. When dry this has the great advantage of being translucent, so that it does not exclude too much light, but a further advantage still is that it can be easily and quickly removed, which is more than can be said for some of the proprietary shading preparations which are used in a similar way.

In hot dry weather humidity can be provided by 'damping down', which means drenching the floor, walls and staging with water, if necessary two or three times a day in hot weather. This, together with the fairly high temperature, will produce the 'growing atmosphere' that we usually associate with greenhouses, and this can be further maintained by standing the pots on a moist bed of sand, peat or shingle. For this purpose an open, slatted staging can be easily covered with asbestos or iron sheeting to provide a base for the sand, etc.

Even in this sort of congenial atmosphere the plants will also need frequent syringing until their flowers start to show colour, when the syringing should be directed at the pots rather than the plants; otherwise the delicate flowers of some of the choice varieties will soon be marked by water. For this syringing a sprayer with a mist nozzle should be used if possible.

Feeding

During the summer, feeding will also be needed. A start should be made as soon as the plants are well established in their final pots, when any of the proprietary fertilisers containing a balanced proportion of nitrogen, phosphorus and potassium should be used according to the maker's instructions, and always when the compost is already moist. A fertiliser of this type will ensure good growth until the plants have reached the desired size, when one containing a rather higher proportion of potassium will help to produce not only more blooms but also better quality ones of a richer colouring.

It is, incidentally, useful to know the different effect that nitrogen and potash have on plants, as this can help to indicate the type of feeding needed. The general effect of nitrogen is to produce strong shoot and leaf growth rather than flowers, but in excess it has a similar effect to that of prolonged wet weather in the open, resulting in soft, lush growth that is susceptible to diseases and other troubles. Potassium, on the other hand, has a similar effect to that of strong sunlight, in that it hardens the growth and assists in the production of plenty of good-quality blooms, although if it is used in excess it can produce small and stunted plants which make little headway. If this happens one or two feeds with a nitrogenous fertiliser will usually effect a remedy. The reverse, of course, applies to plants which are making too much leaf and shoot growth at the expense of the blooms, and which therefore need extra potassium to harden their growth.

Treatment After Flowering

By the end of the summer some of the greenhouse fuchsias will have stopped flowering and they must be gradually dried off as described in Chapter 6 on 'Fuchsias Indoors', although where there are many plants it may not be possible to put them all outside. In this case as much light and air as possible should be admitted to the greenhouse until all the leaves and flowers have fallen from the plants, when they should be gradually dried off and stored as described in Chapter 9 on 'Winter and Spring Treatment'.

8
Training to Shape

Shaping your fuchsia plant into a bush is only one of several different methods of training, some of which have already been mentioned in passing. The 'standard', the 'pyramid', the 'pillar', the 'espalier' and the 'climber' can all be achieved, and there are varieties that may be grown in one or more of these different ways. They all need far more constant attention than the bush form but they are not much more difficult and when successful they are certainly something that you need not be ashamed to boast about. The only snag is that they all require a heated greenhouse, as they cannot be properly grown in one season and consequently have to be kept alive through the winter, when they must be safe from frost.

Standards

The simplest of these forms of fuchsia is the 'standard', which is in effect a bush plant grown on a bare length of stem so that it resembles a miniature tree; and with this it may even be possible to manage without a heated greenhouse. Several specialist fuchsia nurseries offer partly grown standards, known as 'whips', in the spring, and these should make good plants the first year; then after that they can be stored for the winter in any frost-proof place such as a cellar or garage.

The usual height of a standard for greenhouse or outdoor summer use is from 18 to 30in, but there are also small 'table standards', suitable for temporary decorative use indoors, with the main stem no more than a foot long, and these are grown in exactly the same way as the taller ones. Even the 30in ones are not really true standards, as plants considerably taller than this can be grown; but these very tall ones are not usually a proposition for the amateur, as apart from needing a good deal of skill and experience they cannot be conveniently housed in a small greenhouse and unlike the smaller 'half-standards' are not ideal for outdoor summer use; being obviously far more susceptible to wind damage.

The selection of suitable varieties to grow as standards is important and some of those commonly grown in this way are given at the end of this section. These include both those with upright heads and those with trailing or cascade heads, but in general the latter are to be preferred; while the upright ones tend to produce a stiff and formal tree-like shape, these give a graceful 'waterfall' effect that is far more attractive.

It is also necessary to use varieties that do not produce flower-buds too soon. This fault is sometimes due to a check to growth, caused perhaps by a delay in potting-on or neglect with the watering, but even with good cultivation some varieties produce flower-buds at the tip of the stem before the plant has reached the required height. In this case no further upward growth is made and unless you are going to be satisfied with a shorter stem than you anticipated such plants should be discarded. For this reason it is always advisable to aim at rather more standards than you actually need, and to use several different varieties so that if any prove unsatisfactory you will still have sufficient left.

There is one peculiarity of certain fuchsias which indicates their suitability for growing as standards. On most varieties the leaves are produced opposite each other in pairs at each node or joint of the stem, but on some they are produced in threes. It is the varieties with this latter characteristic which usually produce the best standards, as when the main stem is eventually stopped to form the bushy head new sideshoots will appear at each of the three leaves, thus giving three shoots rather than the normal two, and so forming the head more quickly.

When aiming at plants of a certain height it must be remembered that the head will add a foot or so to the height of the stem, so that a plant with a 30in stem will eventually finish at an overall height of 42in. The ultimate height should be particularly borne in mind when the plants are to be grown on a greenhouse staging.

From Spring Cuttings
As early as possible in spring suitable strong-growing cuttings should be selected and grown on in 3 or 3½in pots, each one being tied to a small cane for the time being until a larger one is needed. From then on the main thing is to keep them growing without any check due to too low temperatures, lack of water, or insect pests; then, before the small pots are full of roots, the plants will need potting-on into 5, 6 or 7in pots according to their ultimate size. This potting-on is done in exactly the same way as described under 'Final Potting' in Chapter 7.

At this stage the training of the standard really begins. Before it outgrows the first small cane it must be given another long enough to reach up to where the base of the head will be. Frequent tying, every 2 or 3in, will then be needed to ensure that the stem is kept quite straight and it will also be necessary to turn the plant round, preferably every day, so that it does not grow in one direction only, towards the light. The tying must be firm enough to hold the stem to the cane but not tight enough to strangle it, so the original ties should be replaced with new ones as the occasion demands.

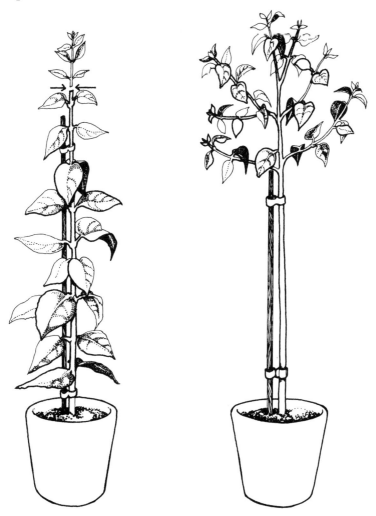

6   Left, a young standard with tip pinched out, side-shoots removed and leaves left on. Right, a well-formed standard after the stem leaves have been removed

On no account must the young plants be stopped. Instead they must be encouraged to grow upwards, with any side-shoots that appear in the axils of the leaves (where each leaf joins the stem) being removed—pinch them out with finger and thumb. It is important to note, however, that it is only the side-shoots and not the leaves themselves that are removed, as the plant will still need these to function properly.

Forming the Head

The removal of the side-shoots continues until the stem has reached the required height, when a start can be made on forming the head. This is done by allowing the stem to continue upwards until a further three sets of leaves have been produced, when the tip of the stem is pinched out (Fig 6). The side-shoots that are produced in these topmost leaves are the ones that will form the branches of the head, but to increase their number they should each be stopped when they have made two pairs of leaves.

An alternative method of training is adopted by some growers. Instead of removing the side-shoots from all the way up the main stem they are merely pinched out to prevent them becoming unduly long; then as soon as the head has been formed they are removed altogether. The theory behind this is that the plant needs all the leaves it can get, both on the actual stem and on the side-shoots, to maintain its strength. But against this it appears that if the side-shoots are removed the actual leaves on the stem grow much larger, thus making up for the leaves lost.

There is, however, something to be said for not removing the side-shoots if the main stem tends to be weak and spindly. By allowing them to grow and eventually stopping them at two or three pairs of leaves the upward growth of the main stem is checked, thus encouraging it to thicken up. Then as soon as the head is formed the side-shoots from the main stem can all be removed, together with the leaves.

From Summer Cuttings

Standards grown from spring-struck cuttings will have started to form the head by the end of the summer, but if a temperature of about 50°F (10°C) can be maintained in the greenhouse better plants can be raised from cuttings taken in summer, from July to September. Potted up in 3in pots before the winter these can then be kept growing slowly right through until the spring, when they should be moved on into 5in pots. Such plants will make good heads by the end of the summer, after which they can be dried off and stored for the winter. Apart from keeping the rooted cuttings growing through the winter the treatment is exactly the same as for standards raised from spring-struck cuttings, but maximum light and minimum watering are needed in winter.

Varieties for Growing as Standards

'Amy Lye'   A Victorian variety with single flowers on a strong, upright plant. The tube and wide sepals are white, with a waxy sheen, and the corolla is cerise-pink. Makes a profuse show of medium-sized flowers.

'Avocet'   Makes a strong upright bush with plenty of bloom. The short tube and the narrow sepals are deep red and the short and plump corolla is white. Makes a good standard.

'Bella Forbes'   An easy grower of vigorous habit. The medium-sized double flowers are produced very freely and are cerise on the tube and sepals and creamy white on the corolla. Makes a fine upright bush or standard.

'Bicentennial'   A very recent introduction with double flowers on a plant of semi-trailing habit. The tube is white, the sepals light orange-pink and the corolla orange, with flaring petals.

'Blue Boy'   Another Victorian variety, and one which is hardy enough to stand most winters in a sheltered position outside. The single flowers, deep pink on the tube and sepals and violet-blue on the corolla, are small but are produced very freely. Makes a strong upright grower. Not to be confused with the modern 'Blue Boy' variety.

'Carole Pugh'   Another recent introduction with double flowers on a self-branching trailer. Can be grown as both a basket plant and standard. The tube and sepals are in two shades of pink and the corolla is amethyst-violet, with a purple base to the petals.

'Dawn Sky'   A large-flowered double variety with the tube and sepals neyron-rose and the corolla heliotrope. The flowers turn purple as they age. There is plenty of bloom on a good upright bush with long leaves.

'Elsa' A semi-double variety that is hardy in a sheltered position. The tube and sepals are pink and the corolla rosy-purple. A very free-flowering plant with fairly large blooms produced early in the season.

'Falling Stars' A fine fuchsia with a profuse display of medium-sized flowers. A vigorous grower of an arching habit. The single flowers are scarlet on the tube and sepals and turkey-red on the corolla.

'Flying Cloud' Almost an ivory-white self, except for a tinge of pink on the tube. It produces plenty of medium-sized double blooms of pleasing shape on a plant of bushy, upright habit.

'Geisha Girl' A fine variety that makes a good bush, standard or pyramid. The profuse double flowers are carried on a self-branching plant and are completely red except for the purple-tinged corolla.

'Jack Acland' A very popular variety, almost a pink self. Makes a profuse display of large flowers on a bushy upright plant. Suitable for both standards and baskets.

'Kathleen' A recent introduction that is good both as a bush and standard. The tube and sepals are red and the corolla orange-white. Self-branching.

'Melody' An extremely free-flowering fuchsia, with single flowers of medium size. The tube and sepals are pale ivory-pink, and the corolla pale purple. Makes a good pyramid as well as a standard.

'Mrs Lovell Swisher' A well-tried fuchsia that makes a vigorous plant with small single flowers that make up in quantity what they lack in size. The tube and sepals are ivory-white and the corolla varies from pink to rose-red. Grow also as a pyramid.

'Mrs Rundle' An old variety that makes a fine weeping standard, with plenty of large, long blooms of graceful shape. The long tube and narrow sepals are flesh-pink and the corolla orange-vermilion. The rather lax growth involves a lot of tying but the plant is worth it. Also makes a good basket.

'Muriel Evans' A single-flowered red self, with medium-sized flowers in profusion against light green foliage. An early-flowering variety forming a strong upright plant.

The pink and white 'Mrs Lovell Swisher', grown here as a bush, is often used as a standard

'Display', a pink single, is an all-rounder – suitable for most forms of training, but equally good indoors or as a summer bedder

'San Mateo' The large double flowers are pink on the tube and sepals and the corolla is dark violet splashed with pink. An upright grower until the weight of blooms causes the stems to cascade.

'Streamliner' A cascade variety with long semi-double flowers, on which the sepals are twisted and curled. Almost a crimson self. Very free-flowering but of a lax habit that needs a lot of tying.

'Ting-a-Ling' A strong upright plant with a very long flowering season. A white self, with the petals of the corolla opening to a bell shape. The medium-sized single flowers are very freely produced.

Pyramids

The growing of pyramid fuchsias, which as the name implies are more or less conical in shape, requires more skill and care than the growing of standards, and the choice of varieties is also more limited. If suitable ones are used—some of these are listed below—and the grower is prepared to pay constant attention to them, good pyramids can be produced in about a year from summer-struck cuttings, or in about eighteen months from spring-struck ones. Such plants will not be like the 10ft giants that the Victorians grew but they should make very showy and shapely specimens up to say 3ft high.

The first essential is a strong, well-rooted and preferably short-jointed cutting of a variety that makes definite upright growth. As with standards, the time to take the cuttings will depend on the heat available in the greenhouse. If a temperature of about 50°F (10°C) can be maintained in winter the cuttings can be taken in August and September and kept growing until the spring; such cuttings should produce good pyramids in the ensuing summer. But if little or no heat is available the cuttings will have to be taken from old plants cut back in spring. They can then be trained to shape by the end of the summer, when they can be dried off and stored through the winter ready for flowering in the following summer.

Training

The actual taking of the cuttings is dealt with in Chapter 10 and as soon as they are rooted they should be potted up into 3½ or 4in pots with a small cane to each one to keep it straight. The plants are kept growing in these pots until they are stopped at about 8in high by pinching out the tip of the stem. This will result in the production of side-shoots, and of these the strongest one near the top should be grown on as a new main stem or 'leader', while the other shoot from the same pair of leaves is removed altogether. All other side-shoots must be retained as these will form the lower branches of the pyramid.

The new 'leader' is then grown on until this again is stopped at about 8in long, or when four pairs of leaves have formed. Of the resultant side-shoots the strongest one near the top is again taken up as a further new 'leader', with its opposite number in the same pair of leaves being removed as before. Again, all other side-shoots must be retained to form further branches of the pyramid.

This same process of forming new leaders and retaining the side-shoots is then repeated until the plant is the required height. The side-shoots on each new leader will obviously be smaller and rather behind those from the previous one, and it is in this way that the pyramid shape is formed.

As the side-shoots grow they are all stopped in turn when they have made three pairs of leaves, when any secondary shoots from them should be reduced to two if there is any sign of overcrowding. The main thing is to make sure that the lowest branches remain the longest. To ensure this they should be trained upwards and outwards, as they will make stronger growth this way than when either horizontal or drooping. Then when the plant comes into bloom the weight of the flowers will bring these lower branches down to a more horizontal position, thus increasing still further the width of the pyramid base.

General Treatment of Pyramids

The main thing in the management of the pyramid is the removal of all flower-buds until the plants have reached their final shape. It may be tempting to leave a few on, just to see what the blooms look like, but the production of even these will put a brake on the growth and that is the very thing that must be avoided.

Delays in potting may also bring about a check to growth. As soon as the roots can be seen in the drainage holes of the pot it is time to move the plants on into larger ones, from the original 3½ or 4in one into a 6 or 7in. Then, when the roots can be seen in the bottom of the final pot, feeding can commence, although this must never be overdone or you will get more leaf than flower when the plant is allowed to bloom. A weak feed about once a week should be ample.

But perhaps the most difficult part is the training. The small cane first provided for the rooted cutting will soon have to be replaced by a taller and stronger one so that the plant can be kept straight all along its length. It may be necessary, too, to provide some means of supporting the lower branches when they are being trained upwards and outwards. This can best be done by providing a 'wigwam' of light canes secured at the top to the central one and with some twine fastened round it to which the branches can be tied. This need only stay in position until the plant comes into bloom, when on its removal the branches will take their correct place.

Varieties for Pyramids

'**Achievement**' The tube and recurving sepals of the single flowers are cerise and the corolla purple. This is an easy and early variety going back to Victorian times. It makes a fine, strong, upright bush that can be trained to almost any shape. It is even hardy outside in a sheltered corner. The fairly large flowers are produced very freely.

'**Brutus**' Another Victorian variety, with single flowers on a plant of vigorous upright habit. The tube and the gracefully recurving sepals are a rich cerise and the corolla a vivid purple. Makes a fine pyramid, enhanced by the dark-veined green foliage.

'**Carmel Blue**' A vigorous grower of upright habit. Very free-flowering with a profusion of single flowers, white on the tube and sepals and an attractive smoky-blue on the long corolla.

'**Checkerboard**' A superb variety with single flowers consisting of a long and narrow red tube with the sepals changing sharply to white where they meet the tube. The corolla is a deeper red than the tube. This variety has all the qualities of a good fuchsia, being vigorous, bushy and early-flowering, with a prolific show of bloom. Makes a wonderful pyramid and is equally successful grown as a standard or bush.

'**Christine Pugh**' A recent variety with medium-sized double flowers on a bushy upright plant of self-branching habit. The tube and sepals are pink and the corolla phlox-purple.

'**Constance**' An easy variety that provides a good initiation into the art of growing pyramids. A strong upright grower with plenty of medium-sized double flowers, pink on the tube and sepals and mauve, tinged with pink, on the corolla.

77

'Fascination' (syn. Emile de Wildeman). An old variety that is still popular as a standard or pyramid and as an exhibition plant. The large double flowers are produced very freely on a vigorous upright plant. The tube and sepals are a good red and the corolla is blush-pink with deeper shading.

'Mrs Minnie Pugh' A recent introduction that can be grown to most shapes. Makes an upright bush of self-branching habit, with semi-double flowers of unusual colouring. The tube is salmon and the sepals are a frosted crimson inside and carmine-rose outside, while the corolla is crimson shading to ruby-red.

'Pink Pearl' Almost a pink self, with the corolla a slightly deeper shade than the tube and sepals. An easy variety to grow as a pyramid, although it also makes a good bush or standard. Makes a profuse display of medium-sized blooms on a vigorous upright plant.

'Royal Purple' An old variety that is excellent as a pyramid, standard or climber. Makes a strong upright plant with plenty of large single flowers of intense colouring, the waxy tube and sepals cerise and the corolla a vivid royal-purple.

'Snowcap' A superb variety for pyramid training. Extremely free-flowering with a splendid display of double flowers, the tube and sepals red and the corolla pure white. Medium-sized flowers on a strong upright plant. Can also be grown as a bush or standard. Sometimes listed as hardy.

'Tennessee Waltz' A very popular variety that is easily grown. Forms a strong upright bush that can be readily trained as either a standard or a pyramid. It has also proved hardy in many areas. The fairly large double flowers are quite prolific and have an attractive colouring, with the tube and upswept sepals rose-madder and the squarish petals of the corolla lilac, splashed pink.

'Violet Gem' An easy fuchsia that makes a good plant for the beginner. The tube and sepals of the semi-double flowers are carmine and the spreading corolla a deep violet. The growth is upright and very bushy, with plenty of large blooms.

Pillars

Unlike the pyramid, the pillar, or 'cordon' as it is sometimes called, consists of a parallel column, which for a true pillar, as they used to be grown, should be about 6ft high and 18in in diameter. Few small greenhouses of today could accommodate a plant of this size without sacrificing too much precious space, but where there is room to grow such a specimen it would make an unusual and very decorative feature for say a patio or balcony, or for the larger greenhouse, conservatory or 'garden room'. The only thing to guard against when such a plant is placed outside is the wind; if you can find a very sheltered corner and perhaps tie it in position it should be safe enough.

But most people would probably prefer a smaller pillar and would be quite satisfied with one say 3 to 4ft high; and to grow such a plant is not difficult. As with the pyramid the first essential is a strong well-rooted cutting of a suitable variety (any of the pyramids are ideal) and this should be started off in the same way and at the same time as the pyramid. The general management, such as potting, watering, feeding and so on, is exactly the same as for the pyramid but the actual training into shape is quite different.

Training

For a small pillar for home decoration, training need consist of no more than allowing the rooted cutting to grow straight upwards, with a strong cane to support it. All the side-shoots from the main stem are then allowed to grow but are kept to a uniform length by judicious pinching out. All flower-buds must be removed until the topmost side-shoots are as long as the lower ones, when the main stem can be pinched out and the plant allowed to flower. It may be necessary to remove some of the secondary side-shoots from the lower branches to make more room for the flowers and also to allow free circulation of air around the plant.

For rather taller pillars, say up to 4 or 5ft, there are two methods of training. One is to grow two rooted cuttings of equal strength in the same pot. These are each kept to one main stem, with one strong cane serving for both, but while on one all the side-shoots are allowed to grow, on the other they are all removed in the same way as for a standard, until the stem is half the ultimate height of the pillar.

The one with the side-shoots left intact will form the lower half of the pillar and should be stopped at half the finished height. The other, with its side-shoots removed, is allowed to go on growing upwards to form the upper half of the pillar, but from a point just above the stopped lower half the side-shoots must be left intact. Thus a continuous run of foliage and side-shoots is ensured, with the upper half taking over where the lower half finishes. The stopping of the lower half checks it sufficiently to give the stem of the upper half time to develop to its full height, when it is finally pinched out and the plant is allowed to flower.

During the whole process all the side-shoots on the plant must be pinched back to a uniform length until the pillar has reached its full height and is starting to flower.

The other method of training is merely a variation on the above. Instead of two rooted cuttings one only is used, but this is stopped at three pairs of leaves. The two strongest of the resultant side-shoots, usually those near the top, are then selected to form the two main stems, and these are then treated in the same way as when two separate plants are being used.

For both methods of training the same varieties are suitable and any of those listed for pyramid culture can be effectively used.

Espalier Fuchsias

The term 'espalier' is applied to a plant trained in one plane, usually flat against a wall, fence or lattice-work, with its branches horizontal. This method of training is generally used for fruit but some ornamentals may also be similarly dealt with and few of these are more effective than the fuchsia.

The most suitable varieties for this method of training are those with lax or flexible branches which can be easily trained into position. Some of the most popular ones are given below.

Methods of Training

In the greenhouse, espaliers may be grown either actually in the ground or in large pots or small tubs. The former method is really only suitable for the lean-to type of greenhouse, where the plants can be grown against the back wall; in a span-roof greenhouse of sufficient size at least one plant can be grown against the end of the structure. This method of growing them actually in the ground produces the largest and finest espaliers, but in the average small greenhouse of today the plants are better grown in pots. Although this involves more watering it ensures perfect drainage, which is not always easily achieved in the actual soil, and it also does away with the need for heat in winter, when the plants can be removed to any frostproof place.

Even when grown in pots there are two ways of dealing with espaliers. One is to grow them as 'free-standing' plants—ones that can be moved about; the other is to grow them on permanent wires fixed to the greenhouse. This latter method has most of the disadvantages that apply to plants grown actually in the soil, although in the winter the plants can be cut loose and transferred to a frostproof place. Grown as free-standing ones, however, the plants can be grown in the greenhouse while they are being trained and then moved outside for the summer when they will make fine adornments for a patio or balcony, or for anywhere else where they can be placed against a wall.

Supporting Espalier Plants
If the plants are to be grown on permanent wires in the greenhouse these must be placed in position well beforehand so that training can start as soon as the young plant is ready. They should run horizontally, with 6 inches between them, and the easiest way to secure them is through 'screw-eyes' placed at suitable intervals.

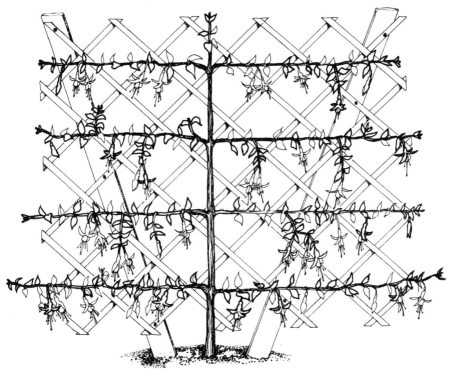

7   Pot-grown espalier trained over a piece of trellis secured to two uprights

For plants grown in pots a suitable framework will have to be erected in the pot itself. Nothing very heavy or strong is needed. Two pieces of 2in × 1in timber firmly inserted in the soil of the pot and splayed out at about 30 degrees will provide a base for a piece of suitable trellis, such as the telescopic or extending wooden type sold at garden and hardware shops; or alternatively the rigid type of plastic-covered wire trellis can be used instead. With either, a piece about 3ft high and say 4 or 5ft wide should be ample for at least the first year (Fig 7).

Potting and Stopping
A start should be made as soon as a strong rooted cutting is available in spring, when it should be grown on in a 3½in pot without any stopping. Then as soon as this small pot is well filled with roots the plant should be moved straight on into a 7in or 8in one, using the JIP No 3 potting compost. The advantage of using this compost instead of a soilless one is that its weight helps to keep the plant standing firmly on its base, and in addition it provides a firm anchorage for the wooden supports of the trellis. This potting is carried out in just the same way as for bush plants, but more care is needed until the roots have taken hold of the relatively large volume of compost.

As soon as this potting is done the plant must be tied to a cane long enough to reach to the top of the trellis and fastened vertically to it. No other canes are necessary, but if you are not using permanent wires some light ones fixed horizontally to the trellis at 6in apart will make it easier to keep the branches straight than when these are fastened direct to the trellis.

The young plant is grown on without any stopping until it reaches the top of the trellis, when its 'growing point' at the tip of the stem is pinched out; or if the plant is making weak growth towards the top of the stem the growing point can be pinched out earlier and a strong side-shoot taken up to form a new 'leader'.

To form the lateral branches, side-shoots corresponding to the horizontal canes or wires are selected and trained out along them, while all other side-shoots are removed. As with the main stem, if any of these lateral branches are making weak terminal growth they can be stopped and other side-shoots taken on in their place as new 'leaders' or extension growths. Stopping of the laterals when they are the required length will result in the production of secondary side-shoots complete with flower-buds along their length, and if necessary these can be thinned out to prevent over-crowding. Flowering should start six to eight weeks after the stopping of the main stem and laterals.

Varieties to Grow as Espaliers
'Achievement'   See varieties for pyramids.
'Carmel Blue'   See varieties for pyramids.

'China Doll'   A fine double-flowered variety with the tube and sepals rich red and the corolla pure white. The large blooms are produced very freely on a plant of cascade habit that makes a superb espalier or basket.

'Golden Marinka'   See hanging-basket varieties.

'Hebe'   A very old variety with a willowy habit that makes it ideal for training as an espalier. The medium-sized single flowers are produced in profusion, with the tube and sepals white and the corolla violet, aging to reddish-crimson.

'Marinka'   See hanging-basket varieties.

'Miss California'   Flower single or semi-double on a lax bush that makes a good basket or espalier. Plenty of medium to large flowers, with the tube and sepals pale pink and the corolla white with touches of pink. An easy variety to train.

'Sierra Blue'   A superb fuchsia of upright habit, with long arching branches that make it ideal for training as an espalier. The double flowers are large and plentiful and their colouring is particularly attractive, with the short tube white and the sepals flushed a very pale pink, making an effective contrast to the full silver-blue corolla, which changes with age to lavender.

'Streamliner'   See varieties for standards.

'Swingtime'   An adaptable variety that may be grown as a bush, standard or espalier. Normally makes a free-branching upright plant with plenty of large beautifully shaped flowers. The tube and short upturned sepals are a rich red and the corolla a milky-white, faintly veined red.

'Thunderbird'   A very free-flowering self-branching trailer, ideal as both a basket plant and espalier. The large double flowers are mainly rose-pink, with the corolla shading to vermilion.

'White Spider' A very useful fuchsia equally good as a bush, weeping standard, espalier or basket plant. The single flowers are almost a white self except for a pink tinge on the sepals. Makes a strong, fast-growing plant.

Climbers

Although none of the countless varieties of hybrid fuchsia is a natural climber, there are some which make superb and unusual plants for training up the roof of a greenhouse, conservatory or glass porch. Here, with their flowers hanging downwards, they can be seen to the best advantage, from underneath; and apart from that they make use of space that would otherwise be wasted. Both upright and cascade varieties can be grown in this way but in the small greenhouse the latter are easier to manage and also provide a more graceful display. Whichever sort is grown the main thing is that it must be a vigorous grower. Enough heat to keep frost at bay is also needed, although in a mild winter the plants will probably be safe enough in a cold greenhouse provided that the roots are adequately protected.

Fuchsias grown as climbers are usually planted either in the actual soil of the greenhouse or in a wooden, brick or concrete 'box' about 2ft each way (trailing plants, including fuchsias, may be used to hide the sides of the box). The John Innes potting compost No 3, over perfect drainage, makes an ideal planting medium, but each spring the top few inches must be replaced with new to make up for the loss of nourishment absorbed by the plant in the previous year. Some wires will also be needed to support the plants. These should run from the eaves to the ridge, and 'vine-training eyes' (obtainable from horticultural sundriesmen) provide the easiest way of supporting the wires a few inches away from the glass.

Planting and Training

A start is best made with a strong, single-stemmed plant as soon as this is ready in spring or early summer. After being planted this is tied to a cane and grown upwards, in much the same manner as a 'standard' fuchsia, with all side-shoots, but not the leaves, being removed. This training continues until the plant reaches the eaves when the plant is carried on up the roof wires. From this point both the main stem and the side-shoots are allowed to develop naturally, with the aim being to reach the required height as quickly as possible. If any check should occur, with a possible weakening of the terminal part of the main stem, the weak part should be cut out and a strong side-shoot taken up in its place as a new 'leader'. The main stem and laterals must of course be kept tied to the wires but the secondary side-shoots will produce the best effect if they are left to develop naturally and hang down gracefully.

During this training of the plant it should be fed with a fertiliser containing a high proportion of nitrogen. This will encourage the growth of the stem and side-shoots at this early stage when flowers are not needed; but when the plant is finally stopped at the required height a more balanced fertiliser should be used. Liberal watering will also be needed, but this should be gradually reduced from September onwards; then during the winter the plant should be kept almost, but not quite, dry until it is started into growth early the following year (see next chapter).

Varieties to Grow as Climbers
'Aztec' A very vigorous plant which makes an excellent climber. The tube and sepals of the double flower are deep red and the corolla violet-purple with small stripes of a brighter red. The large flowers are produced fairly freely against the red-tinged foliage.

'Cascade' See hanging-basket varieties.

'Duchess of Albany' An old single variety with the tube and sepals a creamy waxy-white, with the latter recurving and tinged with pink. The corolla is a pinkish-cerise. The medium-sized flowers are produced on an attractive upright bush that may also be grown as a standard or climber.

'Elizabeth' A fine variety in two shades of pink, with the tube and sepals rose and the corolla salmon. The long single flowers are produced on an upright bush which can be easily trained as a climber. Plenty of 'pinching' is needed whichever way it is grown.

'Fanfare' The tube and small pointed sepals are scarlet and the small corolla is turkey-red. A very floriferous variety of strong upright growth.

'Lady Beth' A beautiful variety with thick crinkly sepals of pale rose and a huge corolla of shining violet-blue. Makes a fine strong upright bush, very free-flowering.

'**Muriel**' Semi-double. A cascade variety with medium to large flowers. The tube and sepals are scarlet and the corolla a bluish-magenta veined pink. Can be readily trained as a climber.

'**Regal**' A very vigorous climber that needs a fairly large greenhouse and is too vigorous for any other form of training. The flowers are semi-double, with the tube, sepals and corolla all rose-madder.

'**Rose of Castile Improved**' Vigorous, pink and purple single.

'**Rose Pillar**' A strong upright grower with large leaves and single flowers of a self-coloured neyron-rose. Makes an attractive climber.

'**Royal Purple**' A single-flowered variety on which the waxy tube and sepals are cerise and the corolla a vivid royal-purple. There are plenty of large flowers on a vigorous upright bush. Can be grown as a bush or standard as well as a climber.

'**Rufus**' See varieties for indoors.
'**Tennessee Waltz**' See pyramid varieties.

9
Winter and Spring Treatment

Like all plants fuchsias need a rest. In the open garden this is brought about by the natural decline in the plants' activity due to the cold conditions of winter; they remain dormant until the more favourable conditions of spring start them into growth again. But when the plants are grown under more artificial conditions, indoors or in the greenhouse, or in containers outside, they must be persuaded to stop growing before the winter as described in Chapter 6; the only point to add here is that at no time must any faded flowers be left on the plants, as they will only weaken them by forming seed. Then by the time all the leaves and flowers have gone the plants will be little more than bare branches, ready for storing away for the winter.

The winter protection of young fuchsia hedges and of the 'near-hardies' has already been dealt with in Chapter 3 and provided that the weather is not too severe there should be no difficulty in keeping these fuchsias alive. But pot-grown plants are a different matter. With these the danger does not lie in snow, frost and the generally clammy conditions of winter: it is dryness at the root which is the main enemy. Experience has shown that no fuchsia, no matter how dormant, can live for long without water; in fact, when hardy or near-hardy fuchsias die in the winter it may well be because in the frozen condition of the soil they just cannot take up all the moisture they need. Not that pot-grown fuchsias need a lot of water when they are dormant: they need just enough to keep them alive without starting them into growth. High temperatures too will start them into growth when they should be resting, and this obviously means that they must be kept as cool as possible.

Indoor Plants

Once these have lost all their leaves they must be brought inside again before there is any danger of a severe frost, although a few degrees is unlikely to harm them. They must not be brought into a heated room, however, as this would tend to start them into growth prematurely. The best place for them is somewhere cool and rather humid, say a cellar, a frostproof garage or garden shed, where if possible they should be laid on their sides to encourage the growth in spring of new shoots from the base. There is no need to cut the plants back yet but for the sake of tidiness and space they may safely be trimmed a little.

A thorough soaking before they are stored away should last them for at least three or four weeks but it is advisable to examine them about once a fortnight so that you can see what sort of condition they are in. A good guide is the state of the tiny pink buds which are waiting to start into growth in spring. If these show signs of shrivelling then the plant is being kept too dry and it should be given more frequent soakings; or if these buds are tending to start into growth the plants are being kept too warm and a cooler place should be found for them. In the necessary cool conditions one thorough soaking with the pot immersed in a bucket of water should last for about a month and half-a-dozen immersions of this sort should see the plants safely through the winter. But do not give these soakings automatically. Examine the plants first to make sure they need them.

Plants in Containers

Perhaps the most difficult plants to deal with are those in hanging baskets or other containers and also those that have been used as bedding plants during the summer. Where these have been merely planted out in their pots there is no difficulty; they may simply be lifted and treated in the same way as the pot-grown ones indoors. But if they have been taken out of their pots and planted straight into the soil of the container or into the soil of the garden they will have to be re-potted before they are stored away.

The plants should first be lifted carefully out of the hanging basket or other container, with the roots intact in the soil as far as possible. They should be placed in pots of an appropriate size and given a thorough watering. They will then need no further watering until they have almost dried out, when after another soaking they will eventually lose all their leaves ready for storing away in the same cool conditions as those grown indoors.

The one exception here is the *triphylla* hybrids which have been used for summer bedding. These are far more tender than the usual hybrid fuchsias and during the winter they must be stored in moderate warmth indoors, or in a heated greenhouse.

Greenhouse Fuchsias

Fuchsias grown as bush plants in pots in the greenhouse can be dried off and stored in the same way as the indoor ones. The most suitable place for them during the winter is beneath the staging of a slightly heated greenhouse in which the temperature can be kept safely above freezing point. Here again they should be laid on their sides—it is well known that when the branches of a shrub are kept in a horizontal position they break into growth far more readily in spring. Standards, pyramids, pillars and espaliers can all be treated in a similar way, although where the standards have been planted out for the summer they will have to be repotted first.

An alternative way of storing pot-grown fuchsias grown in the bush form, and those used for bedding (other than the *triphylla* hybrids), is to bury them in the garden. A sheltered, well drained spot should be chosen and a trench taken out deep enough to bury the plants almost completely. The pots are then stood close together in this and covered with leaves, bracken or ashes before filling the trench in with the excavated soil. This will leave the tops of the plants showing, but as long as the pots containing the roots are at least 6in below the surface the plants are unlikely to come to any harm, particularly if they can also be covered with polythene sheeting.

The most likely cause of trouble with these 'trenched' plants is that the trench may become a sump for water after heavy rain or snow. To avoid this the trench should be on a sloping site so that the lower end can be left open and allow excess water to drain away. If no sloping site is available the bottom of the trench should be lined with rubble to keep the pots clear of any water that may settle in the bottom.

Spring Treatment

Whichever way the plants are stored, by the end of March they will be starting a new lease of life. From then on the new growth must be encouraged as much as possible. It may be that some plants are still so dormant that they might easily be taken for dead. If in doubt, scrape the bark at the bottom of the stem with your finger nail: if green tissue is visible beneath the bark, the plant is alive.

With pot-grown plants stored under cover, the first thing is to place them in the greenhouse or in a sunny window indoors and give them a good watering. In the advancing spring weather the new young shoots will grow away rapidly, and as soon as they are growing well the plants can be pruned. Any weak or badly placed branches should be cut out and the remainder cut back to a suitable shoot near their base, with an eye to ensuring a shapely plant. This will probably leave more new shoots than are required but the surplus ones can be either removed or used as cuttings when they are large enough.

Repotting is the next thing, but this should be left for a week or two after cutting the plant back so that the cut surfaces have a chance to heal; if the plant is cut back and repotted more or less at the same time, there is a chance that with the sudden rush of sap caused by the new compost and increased warmth and water, the cut-back branches will 'bleed', with sap exuding from the cut surfaces.

Plants that have been 'trenched' will probably be rather later in starting into growth, and the covering soil over the pots should be occasionally scraped away to see if the new growth buds are yet visible. If they are, the plants can be removed from the trench and cut back before returning them to it. At this stage there is no need to bury them so deeply, but to guard against late frosts they should be covered completely, with the top of the cut-back branches an inch or two beneath the surface. Then later on, when there is no risk of severe frost, they can be brought out of the trench and repotted.

Repotting

The repotting of the plants is not difficult. They must first be quite moist at the root and then, after they have been knocked out of their pots, as much as possible of the old soil should be removed by poking it away from the roots with a blunt stick. If any of the new white roots are damaged when doing this it will not matter, as the plant will soon form plenty of new ones.

John Innes No 2 is a suitable compost, and the new pots should be the same size as the old ones. 'Crocking', with a few pieces of broken pot placed concave side downwards over the drainage hole is advisable, to ensure perfect drainage, and these should be covered in turn with the rough material left from the sieving of soil and peat.

To repot the plant, stand it on sufficient fresh compost to bring the top of the 'soil-ball'—the soil and roots from the old pot—to about an inch from the top of the new pot; then add more compost to cover the soil-ball, and work it well down around the roots. A few taps on the bench with the pot will then settle the compost down still further, and finally the compost round the edge of the pot should be firmed down with a flat stick so that there are no vacant spaces left. But do not ram

90

'Leonora', an easy fuchsia for indoors, with flowers of a very soft pink

The bushy 'Charming' is an old variety, easy for summer bedding. The flowers are carmine and purple

Popular 'Ballet Girl' does well indoors. Against cerise sepals, the white, fully-double corolla is faintly veined with red

it too hard; it should be just firm enough not to sink when the plant is watered.

When the plants are potted they should be given a good watering and then left until they are almost dry to encourage the roots to take hold of the new compost. Then after another good watering the plants can be grown on normally, with the compost being kept just moist until the plants are growing strongly.

10

Propagation

No plant is easier to propagate than a fuchsia; indeed there is no reason why any fuchsia should ever be completely lost as it can so easily be replaced by a spare one raised from a cutting while the original one is still alive. Such plants will reproduce exactly the characteristics of the parent plant, and it is by this taking of cuttings that the innumerable hybrids are perpetuated.

Fuchsias can also be grown from seed, but unlike plants raised from cuttings they may bear no resemblance to the one from which the seed was taken, they do not come true from seed—in other words they do not necessarily reproduce the characteristics of the parent. Seed is in fact only useful in the hands of the hybridiser (see Chapter 14), who by crossing one variety with another hopes to produce a new and better one. Plants of the fuchsia species—the natural wild forms—do, however, come true from seed, and are sometimes increased this way.

Cuttings are the obvious means of propagation for the home gardener. The only really important point to watch is that each variety is adequately labelled; otherwise there will be no means of telling which is which until they flower, and often not even then!

Cuttings Indoors

There is one very simple way of rooting fuchsia cuttings indoors. This is by merely putting them in a jar of water. From June to September, soft green shoots of the current year's growth, 3 or 4in long, should be taken with a sharp knife or pair of scissors and stripped of their lower leaves. There is no need to trim them off at the base immediately beneath a joint, as is usual with cuttings of most other plants, for the fuchsia roots equally well as an 'internodal cutting', one trimmed off at the base at a point between the joints or 'nodes' from which the leaves spring. Put them in the jar of water, and there they will root, in about three weeks in summer. As they are permanently in water they may be exposed to the full sun, but just in case too much water evaporates, leaving them high and dry, they are safer in a north window, although they may be a little slower in rooting here.

As soon as you can see that the cuttings have made plenty of roots, the young plants are ready for potting-up; use John Innes potting compost No 1 or a soilless potting compost, and 3in or 3½in pots. After one good watering-in they should be allowed to dry a little so that the roots are encouraged to take hold of the new compost; and from then on they may be grown on normally, with the soil kept moist and the plants shaded from the hottest sun.

Although this method is simplicity itself it has two drawbacks. One is that the roots produced in the water are very delicate, so that great care must be taken not to break them off when potting is being done; and the other is that as the plants will not have time to become fully grown by the end of the summer, they will have to be kept growing all through the winter in a temperature of at least 50°F (10°C). But this should not present any difficulty. They will be quite safe in a window indoors, where in an average room temperature they should be kept only just moist enough to keep them slowly on the move.

Spring Cuttings
Cuttings may be similarly rooted in water in the greenhouse during the summer, but here the more usual way is to make use of the young shoots that are produced after the old plants have been cut back in spring. Take these when they have made three pairs of leaves, making the cuts just above the lowest pair, which are left on the old plant to produce a further supply of young shoots. No preparation of the cuttings beyond the removal of the lower leaves is needed, and they can be inserted straightaway into a pot containing a suitable rooting medium.

Fuchsia cuttings will root in almost anything and the rooting medium just mentioned may be sand, vermiculite, coke breeze, a mixture of equal parts sand and peat, or equal parts sand, peat and soil. Personally I prefer this last mixture; all the others are completely sterile and contain no nourishment for the cuttings, which must therefore be removed from the rooting medium before there is any risk of their becoming starved. In the soil, peat and sand mixture there is enough food to keep them going for a time so that if potting-up has to be delayed they will come to no harm. No doubt some growers will say that the addition of soil to the mixture increases the risk of the cuttings being destroyed by the soil-borne fungus that causes 'damping off', but provided that the temperature is adequate (50 to

60°F, 10 to 15°C) there is little risk of this.

Either 3 or 5in pots may be used for the cuttings, but for a comparatively small number the 3in pot is to be preferred. It will take four or five cuttings and by using one pot for each variety it is easy to keep the different varieties separate. Whichever rooting medium is used, make it fairly firm in the pot; then for each cutting make a hole with a pencil-sized dibber near the edge of the pot, insert the cutting and firmly lever up the adjacent soil against the stem by inserting the dibber at an angle and pressing it towards the cutting. The main thing is to make sure that the base of the cutting is firmly in contact with the bottom of the hole; if it is left 'hanging' it might die before it roots. A thorough watering with a fine-rosed can will then settle the cuttings firmly in.

The most suitable place for them depends on the facilities available. The cuttings will root even on the open greenhouse bench as long as they are shaded from the sun and soaked every time they show signs of drying out. Here it may take them up to three weeks to root. Quicker results will be obtained if they can be kept 'close' in a propagating frame; and quicker still if the frame can be heated from underneath (given 'bottom heat'), say by a soil-heating unit, and the air temperature is kept at about 60°F (15°C). Kept warm and close in this way the rooted cuttings will of course have to be 'hardened off' before they are eventually placed on the open greenhouse bench, but this only entails the gradual admission of more air to the frame as soon as new growth indicates that rooting has taken place.

If no propagator is available an alternative method—very useful in the house—is to cover each pot of cuttings with a clear glass jar if a large enough one can be found; or alternatively a plastic bag can be inverted over it and sealed round the pot with a rubber band. It is advisable to remove the jar or bag occasionally to get rid of any condensation, but beyond this the cuttings will need no attention after the first watering-in. Needless to say, they must be stood somewhere out of direct sunlight.

As soon as they are rooted the jar or bag must be removed, and the cuttings must be potted up into 3 or 3½in pots as soon as possible, particularly if they are in a sterile rooting medium. This potting is described in Chapter 7. From then on the young plants should be grown on as described there.

There is, incidentally, little or nothing to be gained by using a hormone rooting powder. Fuchsias root so easily that to do so would merely be a waste of time and money.

Summer Cuttings

The spring cuttings just dealt with will make good flowering plants in the same year and they can then be dried off and stored as 'old plants' for the winter. But this is not so with the summer cuttings taken in the greenhouse from June to September. Like those taken in water indoors they will not have time to develop into mature plants before the winter and must therefore be kept growing then. This will give them a good start over cuttings taken in the following spring, which is particularly important if they are to be trained as standards, pillars, pyramids or espaliers; they can then be at least partly trained by the autumn.

For these summer cuttings unflowered shoots, from 1 to 3in long, of the current year's growth should be taken, and treated in exactly the same way as the spring ones, rooting them either in water or in a rooting medium. Or in the warm summer days another way is to stand the pots of cuttings outside in complete shade, where they will root in about three weeks if kept quite moist. Whichever method is used, no artificial heat is required for any of these summer cuttings. They will root well enough without it.

Apart from supplying useful plants for training, these cuttings will provide plants for early flowering the following year, and in addition of course they provide a good insurance against possible loss of the old plants in winter.

Other Types of Cuttings

Apart from the shoot cuttings just dealt with there are other types of growth that may be used, and for those interested in the techniques of propagation these can be very interesting to try.

One, for instance, is known as the 'soft shoot' or 'bud cutting': young shoots are taken from the cut-back plants in spring when they are little beyond the bud stage, with only two or three embryo leaves showing. These tiny growths are then cut or broken off, the old bud-scales at their base removed and the remaining part inserted in a moist peat and sand mixture in the same warm, humid condition used for normal cuttings; some growers, indeed, do not bother to insert them in the rooting medium but instead just lay them on the surface. In a soil temperature of 65°F (18°C) and an air temperature of 60°F (15°C), rooting takes place in seven to fourteen days, by which time the tiny growths will have grown to almost twice their original size. This method is used by specialist fuchsia growers, who obviously have to produce a large number of plants and who have the necessary equipment to produce the required temperature.

There is another method, of more use to the amateur. This is to use 'leaf-bud' cuttings. These are particularly useful with varieties that make a good number of long branches, as these can be cut into pieces each consisting of one pair of leaves with about ¼in of branch above and below it. This small portion of branch is then cut into two along its length so that each half carries a single leaf. The cuttings are then inserted in a pot of the same compost used for ordinary cuttings, with only the portion of stem buried, the leaf showing above the surface. They are then treated like ordinary cuttings until eventually new growth springs from the dormant bud at the base of the leaf-stalk.

Provided adequate warmth and humidity are available even hardwood cuttings may be used, although they very rarely are. These come from mature branches that have become brown and woody during the summer: terminal pieces a few inches long will then root readily in a cold frame. But they must of course be brought into a heated greenhouse before there is any risk of frost, and grown on through the winter. The old stems removed when the plants are cut back in spring can be similarly rooted in a warm propagating frame, and as already mentioned these will even root in water, although they will take several weeks to do so.

Fuchsias from Seed

Although fuchsias are usually propagated from seed only when hybridisation is being attempted, this method does provide an inexpensive way of acquiring a stock of plants. Mixed seed of some of the best varieties is available, together with that of at least one species, *F. procumbens*. The plants from the mixed seed will be an unknown quantity but they will be sure to include some quite beautiful ones, well worthy of a place in the greenhouse or home.

The seeds are no more difficult to germinate than any of the more common garden seeds, but they have one drawback: they soon lose their viability, which means that they should really be sown as soon as they are harvested (the actual process of harvesting is described in Chapter 14). Obviously this is not possible with purchased seed, which is therefore apt to germinate slowly and erratically, some of the seedlings not appearing for about ten weeks or more. Newly harvested seed, on the other hand, can usually be expected to germinate in three to four weeks. Where fuchsias are grown to any extent, it is not unusual to find stray seedlings coming up in the greenhouse.

For the small amounts of seed that the amateur is likely to sow, one or two seed-pans or pots will be enough. These should be cleaned and well crocked, then filled almost to the rim with John Innes potting compost No 1 or a soilless one. Careful firming of the compost, particularly round the edge, to make it evenly compact throughout is essential, although it should not be over-done with the soilless type of compost. The surface should finally be brought to a smooth, level surface about ½in below the rim, by using either a round wooden 'presser' or the base of another pot.

As the seeds are large enough to handle individually it pays to space them about an inch apart, as this not only avoids the risk of over-crowding the seedlings but also enables them to stay in the seed-pan for a longer period, thus doing away with the usual operation of pricking-out into seed-trays. The seeds should be lightly pressed into the surface and finally covered with no more than ¼in of compost, firmed down gently over them.

Watering is preferably done by soaking the pot from underneath until the moisture seeps through to the surface; the pot should then be allowed to drain before placing it in a temperature of about 60°F (15°C) and covering it with glass and paper. The main thing then is to make sure that the compost does not dry out, and the safest way to avoid this is to stand the pot in a saucer, which should be refilled as soon at it is dry.

Pricking Out and Potting In
As soon as the first seedlings appear the glass and paper coverings must of course be removed, and the seed-pans should be stood in a light position but with shade from full sun. If the seeds were sown individually the young seedlings can be left until they are about an inch high, when they will be ready for transferring straight to 3in pots, although this is an operation that needs care. Each pot should be filled evenly and firmly to about ¼in from the rim with John Innes potting compost No 1 or a soilless compost, and the seedlings are then dibbled in, one to a pot, in much the same way as when inserting cuttings: make a hole deep enough to take the roots, and lever the compost firmly up against them with the dibber. Then after a thorough watering the compost should be allowed to become a little dry, to encourage new root action, before further water is given.

If seeds were sown broadcast rather than individually, they should be pricked out into seed-trays as soon as they can be handled, usually when they are little more than ¼in high. After filling the tray with compost, well firmed down, particularly at the sides and corners, the seedlings should be dibbled in in the same way as for pots, but at 1½in apart each way. Very careful watering-in will be needed to avoid dislodging them. When they have been given a further watering after a period of relative dryness, they should grow away without any trouble until they are about 2in high and ready for their first potting.

One of the *triphylla* hybrids, the rose–pink 'Heinrich Heinkel'

'La Campanella' is an excellent basket variety of bushy habit, with flowers of white and imperial purple

(*left*) The free flowering 'Melody', pink and purple, is good as a bush or standard

(*right*) 'Amy Lye' makes a fine standard with long, white-and-cerise flowers

11

Pests and Diseases

Although fuchsias are relatively free of most pests and diseases, there is one pest that can be a real nuisance—the greenhouse whitefly, dealt with below. Apart from this and, of course, the ubiquitous greenfly, most of the troubles affecting fuchsias are likely to turn up only on occasion and should present no serious problem. With modern chemical sprays, fumigants and aerosols there should in any case be no difficulty in controlling any pests that do appear, *provided that they are dealt with in time.* There is no doubt that prompt treatment is half the battle against them, for while it is easy enough to control a very minor infestation it can be most difficult to deal with one that has got out of hand; and considering the speed with which some pests, such as greenfly, multiply it is usually not long before this unhappy state is reached.

The first and main step towards complete control, in fact, is to keep a careful watch on the plants for the first signs of anything wrong. The underside of the leaves should be particularly watched as it is here that pests can develop unseen, long before their presence is otherwise indicated by distorted, discoloured or sticky leaves. The same is true of diseases, to which the fuchsia is at least as resistant as any other plant. Nevertheless minor troubles do occur occasionally, and as their presence nearly always indicates that there is something wrong with the cultural conditions the remedy is usually simple enough.

With some of the modern insecticides, however, it is possible for the effects of the cure to be worse than the trouble itself. The maker's instructions must therefore be implicitly followed. These may, for instance, indicate certain plants that are likely to be damaged by the chemical and obviously such plants must be removed from the house before any spraying or fumigating is done. Reasonable accuracy in the

This excellent standard is 'Falling Stars', which has flowers in two shades of red

preparation of solutions is also important, for a too weak one may be ineffective while a too strong one may be not only wasteful but positively harmful.

Neither spraying nor fumigation should be carried out in bright sunlight. The best time to apply sprays in summer is in the cool of the evening so that the solution stays on the plants all night. From September to March it is better done in the morning, so that the plants are dry by evening.

Fumigation should preferably be done at night, and the best results will be obtained when the temperature is fairly high, at least 60°F (15°C). Once the operation has been started never stay in the greenhouse for longer than is absolutely necessary; then when you have come out, lock the door and put a notice on it warning anyone that fumigation is being carried out.

One last point; never leave insecticides and similar preparations within the reach of children and never store even the diluted solution in lemonade or similar bottles. To be on the safe side everything should be locked up and clearly labelled.

Pests

Greenfly

There are many different forms of aphides, all closely related, and the commoner of these, such as greenfly, are too well known to need description. Their main effect on plants is to produce a gradual weakening by their sucking of the sap, but in the case of a severe attack other effects may follow. On some of the more susceptible fuchsia varieties it may for instance lead to the yellowing and shedding of the lower leaves, while on others there may be only a certain amount of discoloration and distortion, which in turn helps to produce poor, weak growth. The leaves may become covered with a shiny, sticky substance known as 'honeydew', which is excreted not only by greenfly but also by several other sap-sucking insects. The effect of this is to prevent the leaves from functioning properly and in severe cases it may lead to the growth of a black fungus known as sooty mould (see below).

Fortunately this pest is one of the easiest to control as it is susceptible to many of the more popular insecticides. The systemic ones offer a simple method of control but other sprays or fumigation may be used. Remember that it is essential to direct sprays, including aerosols, on to the undersides as well as the upper surface of the leaves. Fumigation should only be necessary for very severe infestations. The systemic insecticides are the simplest to use on pot-grown plants, and may be applied to the foliage or watered into the soil. Either way, the insecticide is then taken into the sap and subsequently absorbed by the greenfly.

Glasshouse Leafhopper

Although not such a persistent nuisance as greenfly, this pest commonly attacks fuchsias, on which its presence is indicated by a mottling or speckling of the upper sides of the leaves. The adult insects are slender, slightly yellowish creatures about ⅛in long, which are quite active when they are disturbed—hence the name 'hopper'. The eggs are laid in the veins on the undersides of the leaves, where they eventually hatch out into small, almost transparent 'nymphs', which remain more or less inactive as they feed on the sap. As they develop they become more mobile. They finally shed their white moult-skins and emerge as fully active adults. The moult-skins then remain attached to the leaves, where they are often mistaken for actual insects.

Owing to the active nature of this pest, dusting the plants with nicotine (obtainable at garden shops) is the best control, but any of the sprays, aerosols or fumigants used for greenfly should be effective. It will also help if all weeds in the vicinity of the greenhouse are kept down as many of these act as host-plants for leafhoppers.

Greenhouse Whitefly

As just mentioned the white moult-skins of the glasshouse leafhopper retain the shape of the insect when they are sloughed off, and for this reason they are often mistaken for the greenhouse whitefly. This is however a very different pest, resembling more than anything a minute white moth, less than ⅛in across, and just as active; in fact it usually first indicates its presence in the greenhouse by fluttering away in typical moth-like fashion whenever an attacked plant is disturbed.

This is a notorious pest, particularly of tomatoes and fuchsias, although it attacks many other plants as well. It lays its eggs on the underside of the leaves, where in less than a fortnight they hatch out into minute larvae that eventually become more or less scale-like in appearance before they develop into the adult flies. Both the larvae and the adult flies damage the plants by sucking the sap, thus giving the leaves a bleached, mottled appearance, and by excreting the honeydew already mentioned, with the result that the leaves cannot function properly. Sooty Mould (see Diseases below) is likely to develop on this honeydew.

As with greenfly, this pest becomes a problem mainly because of the speed with which it multiplies—in a severely infested greenhouse the flight of the flies can resemble snow. It is most important, therefore, to catch it in its early stages, although even then it is not an easy pest to eliminate. The simplest and best control in the greenhouse is to introduce a minute parasitic Chalcid wasp, *Encarsia formosa*, which is practically invisible to the naked eye. The females of this lay their eggs in the whitefly larvae, which then die and turn black; then finally, when there are no more larvae in which to deposit their eggs, the wasps themselves die out. Supplies of the Chalcid wasp, in the form of parasitised whitefly scales, are available to Fellows of the Royal Horticultural Society when sufficient stocks are in hand, and they may also be obtained from the Greenhouse Research Station, Littlehampton, Sussex. It may also be possible to obtain them from other research stations, particularly if inquiries are made through some horticultural organisation. Dramatic results from its introduction into the greenhouse must not, however, be expected as the control is gradual rather than immediate. The best results will be obtained in summer, when the optimum temperature of at least 70°F (21°C) can be maintained.

Chemical control consists of spraying or fumigating with BHC or malathion, while dichlorvos resin strips may also be used. These insecticides will kill the adult flies but not the larvae, so repeated applications at fortnightly intervals may be needed.

Red Spider

This microscopic pest, smaller than a pin's head, is not actually a spider but a member of another group of 'spinning mites' which make webs similar to those of the true spiders. It can be a serious pest of greenhouse plants, and although the humid conditions needed by fuchsias do not favour it, these plants are still sometimes attacked, particularly where atmospheric conditions are not all they might be. Normally red spider is a serious nuisance only in very hot, dry conditions in summer, and the main preventative is to provide a cool humid atmosphere by spraying, damping down and shading. If even then the pest appears, its presence will soon be indicated by bleached, speckled leaves with a hard, unhealthy look, and by the presence of a very fine webbing in which the eggs are laid on the underside of the leaves. The first batch of these is laid in spring when the bright red females emerge from the cracks and crevices in which they have spent the winter. Further broods follow, but as the summer arrives the females lose their red colouring and both they and the males become more or less straw-coloured.

The pest increases at an enormous rate and control measures must be taken at the first suspicion of its presence. The most effective remedy is fumigation with azobenzine, with a second application ten days after the first, but malathion, petroleum oil and pyrethrum/derris, applied as a spray or aerosol, will give a measure of control.

Vine Weevil

This is an insidious pest if ever there was one, for the first you are likely to know about it is the total collapse of the attacked plant. The only good thing to say about it is that unlike the other pests mentioned so far it does not spread and multiply rapidly, so that perhaps only one or two plants in a batch will be affected.

It is in any case not the weevil itself— a dull black beetle-like creature about one-third of an inch long—that does the actual damage, but the grubs that emerge from the eggs that it lays at the base of the plants during spring and summer. Once a plant has collapsed the only thing to do is to shake the soil off its roots and destroy these grubs, which are about one-third of an inch long and always lying in a characteristically curved position. If there are still a lot of good roots left, the plant should then be repotted into fresh compost to which a little BHC (lindane) dust has been added. In summer, dust this round the base of the plants as a preventative.

Other Pests

Several other pests such as thrips, cyclamen mite, caterpillars and capsid bugs may also attack fuchsias, but none of these should be a serious nuisance and they can all be dealt with by using one or another of the many chemicals available. There is, however, no need to have a whole battery of these, as apart from red spider, which necessitates the use of azobenzine, most pests can be controlled by nicotine, malathion, gamma BHC (lindane) or pyrethrum/derris.

Diseases

Fortunately very few diseases attack the fuchsia. Those that do are nearly always the result of faulty cultivation. To prevent them is therefore not only better but easier than to cure them, although in most cases even attacked plants can be brought back to good health if they are dealt with in time.

Damping Off

This is caused by a soil-borne fungus that attacks the base of the stem and causes it to shrivel and die. It is the one exception as far as the likelihood of bringing the plants back to good health is concerned, for it is nearly always fatal. However, it attacks only seedlings and cuttings, not mature plants. It rarely occurs in the warm, airy conditions of summer, but it can cause serious trouble in the early part of the year, particularly if the plants are kept too moist in a low temperature and with insufficient ventilation. Correct watering, adequate warmth and plenty of fresh air will thus go a long way towards keeping it at bay, but where it repeatedly occurs use either a soilless compost or an ordinary one in which the soil ingredient has been sterilised. A further control preventative is to use Cheshunt Compound, obtainable—with directions for use—at any garden shop.

Seedlings that are attacked are useless but in the case of cuttings it may be possible to remove the top portion above the affected area and re-root this in a fresh, sterile compost.

Grey Mould

This fungus disease, *Botrytis cinerea*, can cause serious trouble to greenhouse plants in the damp atmosphere of autumn and winter. Affected areas will show grey, woolly patches of mildew. It rarely attacks the hardwooded portions of plants but the buds on dormant fuchsias may be attacked and destroyed. Prevention lies in maintaining a dry, buoyant atmosphere with plenty of fresh air and it will also help if extremes of dryness and wetness are avoided. If the disease does put in an appearance, spraying or dusting with sulphur or with any of the modern fungicides will give a certain measure of control, but the best thing undoubtedly is to avoid the cold, damp and stagnant atmosphere that encourages the disease.

Sooty Mould

This fungus disease has already been mentioned as growing on the honeydew excreted by sap-sucking insects, and its name is so apt that there should be no difficulty in recognising it—it looks just like a sooty deposit on the leaves. As it appears only on honeydew the obvious remedy is to keep sap-sucking insects at bay.

Blossom Drop

This is not a disease but a physiological disorder, and as it seldom if ever appears on the hardy outdoor varieties it may be due to either the weaker constitution of the greenhouse varieties or to some fault in their cultivation. The actual cause of the trouble is unknown but it seems likely that under-watering and a too dry atmosphere have something to do with it, so if these are guarded against it should not be a serious nuisance.

Leaf Drop

This again is a physiological disorder, due to some error of cultivation. If it occurs on newly potted plants, over-watering is the most likely cause, but on established ones dryness often results in the shedding of the lower leaves. The use of strong fertilisers, particularly when applied to a dry soil, may also cause serious leaf-fall, owing to the damaging effect on the young roots.

Other Troubles

Hardy fuchsias may also be attacked by the various diseases that trouble other hardy plants, but none of these is likely to be very serious except *Armillaria mellea* and verticillium wilt, both wide-spread in California but much less so in Britain. The former, known variously as honey fungus, bootlace fungus, oak root rot and mushroom root rot, can cause serious damage and eventual death to many garden subjects, including trees, shrubs and herbaceous plants, on all of which its presence is most definitely indicated by the large, honey-coloured fungi which develop in autumn at the base of the affected plant. Another indication of its presence is the white, fan-like sheets of growth found beneath the bark at about ground level, and it is from these that the black, thread-like growths (rhizomorphs) spread out underground to attack other plants.

Where part of a hedge, tree or shrub dies for no apparent reason, the presence of this fungus should be suspected and expert advice sought immediately. There are at least two preparations which claim to be effective against it on the market now and these might well be tried to prevent the trouble from spreading. One is Bray's Emulsion, marketed at the time of writing by Warner Renwick Ltd, PO Box 6, Petworth, West Sussex, and the other Armillatox, supplied by Armillatox Ltd, 44 Town Street, Duffield, Derbyshire.

In Britain, verticillium wilt is more likely to occur on tomatoes and michaelmas daisies than on fuchsias, but where it does occur on the latter it results in the yellowing and eventual death of the lower leaves. However, the disease can be more definitely identified by the brown discolouration visible on the inner tissues when a stem is cut through.

111

As the fungus that causes the disease is carried in the root-stock of the plant it is often possible to obtain a healthy stock of young plants from cuttings taken from the top growths; but unless the variety is particularly scarce it is generally better to burn the affected plants. Grow a fresh stock in a different place where the soil will not be contaminated.

In coastal areas damage to fuchsias may also occur as the result of an excess of salt accumulating in the soil and water. This again is more prevalent in California than it is in Britain. It appears as a scorching of the edges of the leaves. There is not a great deal you can do about it but an occasional thorough watering of the soil should help to wash out the salt, and this operation should then be preferably followed by a thick mulch of moist peat. The use of 'artificial' fertilisers should be avoided as far as possible, the supply of nitrogen in the soil being kept up by the application of organic ones, such as hoof and horn, dried blood and natural manure.

The popular 'Citation', which has large flowers in profusion, forms an upright, well-branched bush or can be grown as a standard *(Harry Smith)*

12

Exhibiting

The general revival of interest in the fuchsia since the last war has led to an enormous increase in its use as an exhibition flower, for which its innumerable varieties and various methods of training make it ideal. In both Britain and the USA magnificent displays are put on by the fuchsia societies, and it is indeed only by seeing these and the equally splendid trade exhibits at the major flower shows that one can come to appreciate the full potentialities of this fascinating plant.

To the beginner who sees such superb specimens it may seem impossible ever to compete with them, but this is far from the case. There is nothing magical or mysterious about their production—just a sound knowledge of the plants' requirements combined with dedication, patience and infinite care and attention down to the most minute detail. There is no obstacle in the way of the enthusiast who wishes to grow his plants to the highest pitch of perfection and there is no surer way of achieving this than via the competitive show bench, which will provide not only a stimulant and incentive but also a standard by which to assess the quality of his own plants.

With this aim in mind the first step should be to join a fuchsia society, as apart from the many friendly contacts it will bring it will provide a fertile source of help and information, and also lead to the exchange of varieties with other members, to the mutual advantage of both. The British Fuchsia Society (formerly The Fuchsia Society) is the leading organisation in this field in Britain, while in the USA the American Fuchsia Society holds a similar position. The main shows held by these organisations are physically out of reach for many people, but in most areas there are smaller societies, generally affiliated to the main ones, which hold shows either on their own or in

Hanging baskets bring colour to any space—backyard, patio, balcony or porch. For best results use fuchsias alone, preferably in one variety like this basket of 'Cascade' (Michael Warren)

conjunction with the local horticultural society; and even the smallest local. flower show is not to be despised, as many an exhibitor at the main shows has started on this very modest scale. To win a prize at a small show against comparatively weak opposition may seem trivial, but at least it is a step towards bigger things!

The Schedule

There is one point on which many newcomers to exhibiting fall down, and it has nothing to do with the growing of the exhibits. It is merely that they do not take sufficient care in studying the schedule, so that no matter how good their plants may be they stand no chance simply because they do not conform to the particular class in which they have been entered. The organisers of shows go to great trouble to prepare these schedules and it is hardly surprising, therefore, that they expect the exhibitors to stick strictly to them.

When you get the schedule make sure that you read it very carefully and note every detail about the classes you intend to enter. Note such things as the permitted size of pot, the size of baskets and the number of plants allowed in each, whether the plants are to be single or double-flowered, and whether one or more varieties are required; and in the case of cut flowers and flower arrangements make sure that your exhibit contains no more and no less than the number stipulated.

Particular note should be taken of the requirements for standards and pyramids. In Britain, for instance, table standards should be from 10 to 17in, half-standards 18 to 19in, and full standards 30 to 42in; and for exhibition, pillar fuchsias are invariably of the single plant type.

Preparing for the Show

Having decided on the classes you intend to enter, make sure that you have more plants available than you actually need; then if something goes wrong with any of them you will still have some to choose from. Give each plenty of room and aim at getting them to the peak of perfection just in time for the show. Obviously this is not very easy for the beginner, but it will help if you understand that a shoot showing green flower-buds will, if pinched out (thus removing these flower-buds), produce side-shoots that will flower about six weeks later; and that it is generally advisable to allow a few more days than this—it is better to be early rather than late with the blooms, as once the flower has formed it will stay in perfect condition for at least a few days. To preserve the shape of the plant it may also be necessary to pinch out those shoots that are not already showing buds, but in this case about eight weeks should be allowed.

Staking of the plants, when needed, must of course be done as discreetly as possible, using adequately strong canes and either raffia or soft green garden twine. It may indeed be advisable to renew the canes and ties just before the show so that they have a clean neat appearance, but this must only be done if it can be carried out safely, with no risk to the stems or blooms. The staking and tying of fuchsias is in fact quite an art, and the beginner will do well to study the prize-winning exhibits at the shows to see how it should be done—this is one of those little but all-important items that can be so easily overlooked when you are lost in admiration of the exhibit.

Before the show, remove any faulty blooms and disfigured leaves, and make sure that the pots and compost are as clean and neat as you can make them. Plastic pots have the advantage here as they can be so easily made like new, but a good scrubbing with wire wool will usually make clay ones quite presentable; or if they are very dirty it may pay to transfer the plant to a new pot, although this can be a little risky. The surface of the compost should be freshened up too, either by re-surfacing it with a little new compost or by top-dressing it with peat or moss.

Lastly, make sure that each variety is neatly, clearly and correctly labelled. A wrongly named plant will not stand a chance. Hanging baskets, by the way, should have the label suspended beneath them.

Transporting the Plants
The evening before the show give the plants a thorough watering so that they will have drained by the time they are to be transported to the show, which can be something of a problem unless you live near enough to be able to carry them there. Otherwise some form of transport will obviously be needed and a private car is not always suitable for such awkward things as standards and hanging baskets. However, if only one or two of these difficult items are to be carried it will be possible to find some obliging passengers to hold them so that they come to no harm. Bush plants can usually be carried in the boot or trunk, although even here they will have to be securely wedged, or packed in a box, to prevent them from falling over.

For a greater number of plants a roomier vehicle will probably be needed, together with some means of keeping them upright. One of the best ways is to use a large, shallow box with a number of circular holes, rather smaller in diameter than the tops of the pots, cut in the bottom of it. If this is then inverted, each plant, even if it is a standard, can be dropped into a hole, in which it will be perfectly stable.

Hanging baskets can be very awkward things to carry but here again good use can be made of a box, this time used the right way up. An upright post is then fitted to each end, to support a horizontal bar in which a hole has been drilled to take the hook of the basket, which will then swing perfectly safely above the box. Another and simpler method is to use a clean oil drum half filled with sand or soil to keep it steady, so that the basket can sit securely in the top of it.

What the Judges Look For

The judging at shows must be to some extent subjective, so that unless an exhibit is so perfect that there can be no doubt about it, one judge may differ from another in his assessment of it. It is not impossible, for instance, for one judge to place more emphasis than another on a certain aspect of the exhibit, and although in practice the system works well enough it is to avoid such situations as this that at the time of writing consideration is being given by the American Fuchsia Society to an alternative method of judging, on a points system. By the time this book appears, in fact, this may have already been adopted in America, perhaps to be followed by a similar arrangement in Britain.

Presumably such a system will be largely based on the qualities that judges already look for, one of the first of these being the general presentation of the plant or plants. Obviously if an exhibit is not attractive as a whole it will stand little chance of receiving an award, no matter how good it may be in some respects. On the other hand an exhibit that has the basic qualities of shapeliness, neat and unobtrusive staking and a well-cared for look (including the pot and compost) will at least lead the judges on to look for further good points.

Among these, of course, will be the condition of the blooms, which should be as near perfect as possible even if this means removing one or two that show any signs of a blemish. The size and number of flowers will also be considered, but in this case the standard by which the plants will be judged may vary widely according to the variety (which emphasises the need for correct labelling). Thus, while one plant may be a perfect example of its kind with only a few very large blooms, another may be at the peak of perfection with far smaller flowers but a greater profusion of them.

Clean, healthy foliage is another thing the judges will look for. Any exhibitor will have the sense to remove any leaves that have obviously been disfigured by insect pests, but care should also be taken to clean off dust and also any residue left by insecticides. Even though the leaves may be fairly old they should look like new.

When preparing a plant for exhibition it is in fact a good idea to put yourself in the position of the judge, and look for every possible fault that you can find; for if you can find it, you can be sure he will. And finally, find out all you can about the varieties that you intend to show, so that you will know what they are capable of. As long as you grow varieties that are likely to catch the judges' eye there is little to gain from changing them each year. Far better to be able to judge them for yourself in comparison with those of previous years.

13

Some Fuchsia Species

Some of the *species*—that is the different natural forms of the fuchsia, as distinct from those bred in cultivation—make interesting and unusual additions to a fuchsia collection. Some of them do not of course stand comparison with the highly bred varieties familiar today, but if you grow them you do have a double satisfaction: firstly in growing something out of the ordinary, and secondly in knowing that you are helping to save these plants from oblivion, a fate that seems likely to befall many of the less common forms, not only of fuchsias but of many other plants as well. Many of these fuchsia species are in fact now grown only by a mere handful of enthusiasts, whether amateur or commercial.

The following list includes the most suitable species to try. If there should be any difficulty in obtaining them, the problem can usually be solved by becoming a member of the Fuchsia Society, who are more than anyone else in touch with possible sources.

The known species not included here are by and large of little interest to most garden fuchsia growers. Even if obtainable, it is unlikely they would be grown to any extent, as many of them are hardly garden-worthy.

F. alpestris   see *F. regia*

F. arborescens   Although rarely listed in nursery catalogues this species from Mexico is worth mentioning here, as it is something of a curiosity. Unlike other fuchsias it produces its small, mauve-pink flowers in terminal spikes resembling those of the lilac (it was originally known as *F. syringaeflora*, from the botanical name of the lilac, *Syringa*) and its leaves also differ from the usual type in being quite large and oval. Possibly one reason for its scarcity is its size, for it eventually forms a small tree or large shrub up to 20ft high. Apart from this it could make a most interesting plant for a cool greenhouse, particularly as it flowers in winter, from October to February.

F. boliviana Like the previous one this species from Bolivia deserves to be better known, although here again its size is a disadvantage as in a heated greenhouse it will form a quite large shrub. The long, deep red flowers, set against softly hairy leaves, are produced in drooping clusters, and are followed by large, edible berries. It is obtainable from a few specialist growers.

F. cinnabarina see *F. reflexa*

F. coccinea From a historical point of view this Brazilian species is interesting in that it was the first fuchsia to be introduced into Britain. Apart from this it has little more to offer than its better-known Chilean relative *F. magellanica*, which it closely resembles except for its longer flowers, downy instead of hairless shoots, and smaller size. It is also just as hardy and as it seldom exceeds 3ft or so in height it makes a good substitute for *F. magellanica* where space is limited.

F. cordifolia This Mexican species has unusual colouring for a fuchsia, with the tube being red, the sepals red and green, and the corolla wholly red. It is not hardy enough to grow outside but is quite safe in a cool greenhouse, where it should eventually form a somewhat ungainly plant 3 to 4ft high, with almost heart-shaped, slender-pointed leaves arranged oppositely or in threes.

F. corymbiflora In many respects much like *F. boliviana* mentioned above, but this fine Peruvian species is of a less branching habit, so that where there is room it may be effectively grown as a greenhouse climber. It forms a very strong-growing plant with large terminal clusters of long, bright red flowers; there is also an even better form of it in the variety *F.c. alba*, on which only the corolla is red and the tube and sepals white.

F. cottinghamii Hardy enough to be grown outside in mild areas, this is a rather taller version of the better known *F. encliandra* (see below) but with smaller flowers. These are red, with reflexing sepals, and they are followed by bead-like fruits of a glossy purple-brown.

F. encliandra (syn *F. parviflora*): Similarly hardy enough to be grown outside in mild areas, this Mexican plant forms a rather rampant dwarf bush with minute red flowers. The leaves too are small, giving the whole plant a somewhat fern-like appearance.

121

F. excorticata   Unlike most fuchsias this is a native of New Zealand, where it forms a small tree, with loose, papery bark and brittle branches. In the mild south-west of England it has been grown to similar proportions, but in the other southern counties it is not likely to make more than a small bush and elsewhere it cannot be relied upon outside at all. The flowers, which are unusual in that they are produced in spring, are up to 1¼in long and in the initial stages green; then as they develop, the tube and sepals become purplish-red and the corolla a very deep purple, against which the blue pollen and the yellow stigma show up well.

F. fulgens   Of Mexican origin, this is one of the finest of all the fuchsia species, forming a tender shrub about 4ft high with long-tubed, bright red flowers tipped with green on the sepals. It is one of the ancestors of many of the greenhouse varieties but is also well worth growing in its own right.

F. gracilis   A fine form of *F. magellanica* and equally hardy. Makes a beautiful very free-flowering shrub of slender arching habit, with small flowers, scarlet on the sepals and violet on the corolla.

F. magellanica   see Fuchsia Hedges, Chapter 3.

F. microphylla   'Microphylla' means small-leaved and this indicates the most characteristic feature of this Mexican plant, as the crowded leaves are no more than ¾in long. The flowers themselves, deep red with a rose-red corolla, are equally tiny, but although they are produced only singly there are usually enough of them to make a quite effective display. It is not hardy enough to grow outside in Britain but in a cool greenhouse it will make a plant up to 6ft high, flowering in the autumn.

F. minimiflora   In this case the specific name means small-flowered, and there could hardly be an apter one for this, the smallest-flowered of all fuchsias, with blooms little more than ⅛in long, white flushed red. Of Mexican origin it is perhaps more of a curiosity than anything else and for this reason would make an interesting addition to a greenhouse collection.

F. parviflora   see *F. encliandra*

F. procumbens   see Fuchsias in the Rock-Garden, Chapter 3.

F. reflexa (syn *F. cinnabarina*)   A similar small-leaved plant to *F. encliandra* but with cerise flowers followed by black fruits.

F. regia   Also known as *F. alpestris* this is a Brazilian plant of more or less climbing habit, with soft, downy leaves and small red flowers produced singly in the leaf-axils at the ends of the shoots. It is almost hardy but is safest in a cool greenhouse, where it will eventually make quite a large plant.

F. simplicicaulis   A Peruvian species which in nature forms a largish shrub but which in the cool greenhouse is better grown as a climber. It is a striking plant, with showy clusters of long, bright red flowers each up to 3in long, and a rather unusual feature are the long, narrow leaves, arranged in whorls of four at intervals along the stem. It is a late-flowering kind, seldom producing its flowers before October.

F. triphylla   As mentioned in Chapter 15, this species native to Haiti and San Domingo was the first of the whole family to be discovered, but although it is still grown it has largely been superseded by the '*triphylla* hybrids' mostly bred from it. Nevertheless, the species. itself makes a very handsome plant, of erect, bushy habit with terminal clusters of vivid red flowers, each of which consists mainly of a long tube. Unlike those of most fuchsias the stamens do not protrude beyond the insignificant corolla. It is one of the tenderest of all fuchsias, but it is safe enough in a cool greenhouse and it would make an interesting addition to any collection.

14

Hybridising

To most people the pleasure obtained from the growing, and perhaps exhibiting, of fuchsias will be sufficient in itself. No doubt, however, there will be others of a more experimental turn of mind who would like to try hybridising, or crossing one variety with another.

There is indeed no reason why the amateur grower should not attempt this fascinating process, for the only equipment required consists of a pair of sharp, pointed scissors (the type used for grape-thinning is ideal), a few small muslin bags—and a lot of patience! It is of course unlikely that anything startling, or even different, from the thousands of varieties that are already available will result, at least from the first efforts. Nevertheless, even to attempt the art, or science, of hybridising will lead to a far greater understanding not only of the fuchsia but of all those plants, common enough in gardens, that are the result of the hybridiser's skill.

Making a Start

It is usually said that when hybridisation is attempted the first thing is to know what you are aiming at—such as a new colour or combination of colours, a different shape of flower, plants of a different habit or of hardier constitution. However, although this is all very well for the more advanced worker it is a rather lofty objective for the beginner, who obviously must first master the basic elements of the operation; and this can be learnt from any two different varieties of fuchsia as long as the plants have some blooms at the proper stage of development.

The flowers to use should be chosen from the first ones to be produced on the selected parent plants, which are then at their greatest vigour. To understand when they are ready, indeed to understand the whole business of hybridisation, some knowledge of the construction of the flower is essential. As this has already been dealt with in Chapter 2, there is no need to elaborate upon it now, beyond saying that, as described there, each flower, if left to itself, would become self-pollinated. This is of course the very thing that the hybridiser wishes to avoid, and to ensure that it does not happen certain precautions have to be taken.

The first thing is to select a bloom on one of the parent plants to act as the female. For this purpose choose a bud that is just on the point of bursting and gently 'pop' it open, between the finger and thumb, so that it reveals the corolla, stamens and stigma. Then turn back the sepals, and if necessary the petals forming the corolla, so that the anthers can be cut from the stamens with the scissors, thus leaving only the projecting style with the sticky stigma at its tip. The bloom should then be immediately totally enclosed in a muslin bag to prevent accidental pollination by bees or other insects.

Once this stage has been reached there is no hurry to pollinate the 'female' flower, and the selection of a suitable flower on the other parent plant to act as the male may if necessary be left for a day or two. The main thing is to choose one on which the pollen is quite ripe: this can be readily ascertained by tapping the stamens with the finger nail, to which some of the pollen will adhere as a powder if it is in the right condition. In some of the highly-bred doubles, by the way, the stamens may not develop properly, but if the plant is allowed to become a little starved it will probably produce flowers with perfect stamens before long.

To effect the actual pollination, remove this male bloom and carry it carefully to the female one, taking care not to dislodge the pollen. Remove the protective bag from the female and bring the anthers of the male bloom into contact with the stigma of the female one, so that the pollen is transferred from the one to the other. It is quite easy to see when the stigma is well dusted with pollen, and immediately this has been done the muslin bag should be replaced. The male bloom can then be discarded.

More than one flower on each plant can of course be pollinated in this way, but with some of them it is just as well to reverse the order of things, using the plant that first provided the male flower to provide a female one, and vice-versa.

Finally, make sure that each pollinated bloom is securely labelled with the date and the names of the parent plants; or alternatively it may be merely given a number, which with the appropriate details is then recorded in a notebook—this is actually the better way as you will then have a permanent record of this and other future crossings. It is usual to place the name of the female parent first in such records.

After Pollination

Once pollination has been made the next step lies with nature. While it is in contact with the stigma each pollen grain sends out a 'pollen tube', which makes its way down the style and into the ovary; there it releases the male sperm cells ('gametes'), which unite with female reproductive cells, thus forming the embryo seed. As soon as this takes place the fertilised flower fades, leaving the seeds enclosed in the berry. At this stage the dead flower should be removed but it is just as well to replace the protective muslin bag, as this will then catch the berry if it happens to ripen and fall off unexpectedly.

When the ripe berry has been gathered the next thing is to separate the seeds from the fleshy pulp forming the outer casing. This is a rather messy business, particularly as the berry can sometimes be quite large and soft, but if the outer casing is first cut into segments with a knife it should not be too difficult to find the seeds by pressing the pulp with the tip of the finger until they can be felt. They should be carefully washed and dried, then either sown or stored away in sealed polythene bags for the winter. But do not forget to label the bags!

Ideally the seed should be sown as soon as it is harvested in late summer, for as pointed out in Chapter 10 it is likely to yield a far better and quicker germination then than if it is kept until spring. The plants from a summer sowing will also flower much earlier in the following year than those from a spring one, and this obviously means that there will not be such a long wait before the quality of the seedling plants can be assessed. As I have already said, there is not likely to be anything very striking among them—this would be most unusual in a first cross; nevertheless, even if they turn out to be inferior to the parent plants they should not be discarded. They should in turn be crossed among themselves, as this second crossing may well produce something worthwhile. This is because in all plants there are certain characteristics termed 'dominant' which are likely to appear in the first cross, and others, termed 'recessive', which may not appear until the second cross. Thus even though the results of the first experiment may be a bit disappointing there is always the excitement of wondering what will turn up when the process is repeated.

Going Further with Hybridisation

These last observations are of course based on the Mendelian Laws, propounded last century by Gregor Mendel, an Austrian monk who first showed that heredity is not a capricious thing producing results at random but something that largely follows certain simple statistical laws. To pursue this subject further is far beyond the scope of this book, and for those who wish to carry the process of hybridisation beyond the very simple outline given above, a considerable knowledge of the genetic factors involved is necessary. There is no shortage of literature on this fascinating subject, but the three books listed below should be more than enough to provide a good start. And apart from that they will certainly present an aspect of our cultivated plants of which the average gardener has barely the faintest inkling.

Practical Plant Breeding, by W. J. C. Lawrence (Allen & Unwin, 1951)
The Genetics of Garden Plants, by M. B. Crane & W. J. C. Lawrence (Macmillan, 1947)
Methods of Plant Breeding, by H. K. Hayes, F. R. Immer and D. C. Smith (McGraw Hill, 2nd edition, 1953)

15

Origin and History

The fuchsia takes its name from Dr Leonard Fuchs, a leading German physician, herbalist and botanical artist of the sixteenth century, although actually he had nothing to do with the plant and had never even heard of it. He was fortunate enough to have his name perpetuated in this way by a botanically-minded French missionary, Father Charles Plumier, who not only collected and recorded many of the plants of the then 'New World' of America but also had a habit of naming his new discoveries after famous botanists of the past. The plant he named 'Fuchsia' is now known to have been the species *F. triphylla*, which he discovered in San Domingo in 1703 and which he illustrated, rather crudely, in his book *Nova Plantarum Americanarum Genera*—the first record of a genus which was eventually to provide us with some of our favourite and most useful plants.

After this somewhat inauspicious start little more was heard of the fuchsia, except perhaps in learned botanical circles, for the better part of a century. It was not until 1788 that the first species was introduced into Britain, under rather peculiar circumstances. The story goes that in that year a plant of the Brazilian *F. coccinea* was given by a Captain Firth to Kew Gardens, while at almost the same time a nurseryman of Wapping, James Lee, gained possession of an identical plant, which he claimed to have bought for 8 guineas (a considerable sum in those days) from a widow whose son had brought it back from South America. This may or may not have been true—his story has been rather discounted recently, and the suspicion seems to be growing that he obtained his plant from the one at Kew under fairly dubious circumstances. After all, most of us have pinched a cutting from somewhere or other at some time, and this is all he would have had to do to obtain a flowering plant in the same year, if the cutting was taken in spring. But if this was what he did it was certainly a profitable bit of pilfering, for by 1793 he was selling similar plants at from 10 to 20 guineas each!

During the following forty years or so most of the species that were to prove of any value were introduced. It was not these themselves, however, but the hybrids developed from them that were to attract the attention of the gardening world. One species did prove invaluable, both as a parent of the new hybrids and as a garden plant in its own right, and that was *F. magellanica* from the southernmost part of the American continent. In that area the weather is certainly no better, and often worse, than in the British Isles, and thus, once it was introduced, this species soon made itself at home even as an outdoor plant; and indeed it is still this one, together with its sub-species and varieties, which now forms the bulk of the fuchsia hedges that are such a noticeable feature of the milder parts of Britain, particularly south-west England, western Scotland, and of Co Kerry in Eire.

Cultivation in Victorian Times

Unfortunately the absolute hardiness of this particular species was not transferred to its offspring, and although the new hybrids were vastly superior in the size, colour and shape of their blooms they were also too tender to be grown permanently outside. It was, therefore, primarily as greenhouse plants that they were developed, and it was in fact in this particular field that they were to take the gardening world by storm; during the latter half of the nineteenth century the cultivation of fuchsias became almost a craze.

At that time there were of course none of the horticultural aids that we have today in the way of efficient heating, pesticides, fungicides, fertilisers and so on (to say nothing of a vastly superior theoretical knowledge). But labour was cheap, time of little account and distractions in the form of travel and entertainment almost non-existent. Time, work and in many cases money were therefore lavished on the popular plants of the time, and the fuchsia eventually reached a pitch of perfection that it has never reached since and probably never will again. (It is said that on some of the large estates gardeners were employed only to watch the specimen fuchsias right round the clock, so that if a faulty leaf or misplaced shoot showed up, even in the middle of the night, it was immediately either removed or corrected!)

With such attention it is hardly surprising that the results were amazing. It was not unusual, for instance, for plants to be grown in the form of standards, pyramids or pillars up to 10ft tall; the latter two forms would be clothed from head to foot with blooms. Equally superb specimens were grown in baskets, as espaliers—plants trained out flat against a wall—and also as climbers that reached to the top of the greenhouse roof. In most cases the same varieties were used over and over again for these pampered plants, so that their habit of growth could be anticipated; but at the same time, as today, literally hundreds of new varieties were being introduced. Some of these are with us still, others had their brief day and disappeared, as they could not meet the stringent demands of the time.

Recent History

With the coming of the First World War the prodigal expenditure of time, labour and money on plants came to an end. But why for the next twenty years or so the fuchsia almost lapsed into oblivion is anybody's guess, for more modest specimens can be grown even on a shoestring expenditure. Possible reasons were the fashionable revulsion against all things Victorian, the hard times which led to the growing of tomatoes and other edible crops in the greenhouses, and the fact that just simply the plants went out of fashion, for flowers, like many other things, come and go with the times.

But even during this period in the doldrums there were a few enthusiasts who believed that all was not lost. A handful of these were in Britain, but it was in California that the keenest interest was shown, probably because the climate there, particularly in the Bay area of San Francisco, with its moist air and summer fogs, is so ideal for fuchsias that it is not necessary to cultivate even the tender forms under glass. Here, in 1929, the American Fuchsia Society was founded, and with that the fuchsia started on the road back. Great efforts were made by its members to rescue the surviving varieties that were still in existence in Britain and Europe, and as an example of this one member, Sydney B. Mitchell, managed to send home fifty different varieties, some of which are still in cultivation in America today. It is interesting to note, by the way, that when the Victorian craze for fuchsias was at its height, the plant was comparatively unknown in the USA, with practically no nursery listing more than about half a dozen varieties.

The medium-sized flowers of 'Bella Forbes' are produced freely, making a fine upright bush or standard *(Harry Smith)*

'Iced Champagne' is an old variety which makes a good pot-plant. Such smaller uprights are also ideal for window boxes, behind a trailing variety *(Harry Smith)*

At the same time as these old varieties were being gathered together, intensive hybridisation was being carried out. Although at first the resultant new varieties were circulated only fairly locally, mainly in the Pacific States of California, Washington and Oregon, their influence was eventually felt in Britain, where in 1938 the Fuchsia Society was founded. Eventually to be known as the British Fuchsia Society this numbered among its members many dedicated enthusiasts who devoted their skill and knowledge to the furtherance of their chosen plant, so that nowadays, although many of the new varieties still come from America there is an increasing number of British ones. Others are coming from the Continent, and in Germany particularly there has been an increasing interest in a different range of varieties known as the *triphylla* hybrids, largely developed from the same species that was the first to be discovered in America.

The ever-increasing number of new varieties is not, however, without its drawbacks, the main one being the confusion of names. In some cases the same name has been applied to different plants, while in others the same plant has been given different names. One of the first efforts to sort these out was made in 1936 by Dr E. O. Essig of the University of California, who in his *Check List of Fuchsias* listed over 1,900 different names—and that was before the real flood of modern varieties had started! This is of course now long out-of-date, but it still provides the basis for more modern lists. The greatest step towards clarification has been the appointment of the American Fuchsia Society as the International Registration Authority for Fuchsias, an appointment made by the International Society for Horticultural Science in 1967. If it does nothing else it should at least ensure that old varieties do not reappear under new names, a state of affairs which has largely been responsible for the confusion in the past.

Three fairly hardy, adaptable varieties are used here in baskets to brighten up a fence: 'Lena' (white and purple), 'Swingtime' (red and white), and 'Tennessee Waltz' (red and lilac) *(Michael Warren)*

The Fuchsia-Grower's Calendar

Owing to variations in the weather year by year and in different areas, and in the amount of heat available in the greenhouse—if any—it is not possible to give exact dates when the various operations should be carried out; but the following should prove helpful as a reminder list.

January

1 Where young fuchsia hedges and established 'near-hardies' have not yet been given their winter protection, get this done immediately—see November.
2 Continue to examine stored plants to make sure they are not too dry—see December.
3 Take precautions against mice and woodlice, if these attack stored plants.
4 Summer-struck cuttings potted before the winter: these need enough water to keep them growing in a temperature of about 50°F (10°C). Keep them in full light.
5 Check all show equipment, window-boxes, troughs, etc, and repair where necessary.
6 Send for catalogues and order new varieties for delivery at the appropriate time. At the same time order 'whips' for growing on as standards.

February

1 In a heated greenhouse start old plants into growth, from the middle of the month onwards, by giving them more water and spraying them overhead in the mornings of bright days. Cut them back as soon as new growth can be seen.
2 Continue to keep summer-struck cuttings on the move, with warmth and water. Those to be grown on as bush plants can be stopped as soon as they have made three or four pairs of leaves.

3 Continue to check stored plants for dryness, and remove all weeds, fallen leaves and other debris from their vicinity to reduce the risk of slugs and other pests.

4 As soon as the summer-struck cuttings that were potted earlier start to root through the drainage holes, move them on into 5 or 6in pots. Water sparingly at first, shade them from the sun and maintain a temperature of 45° to 50°F (7° to 10°C).

5 Prepare a supply of compost and clean pots ready for the main potting time later on.

March

1 Repot old greenhouse plants a few days after they have been cut back. Water the repotted plants sparingly until new growth indicates that root action has taken place and keep them in a temperature of 45° to 50°F (7° to 10°C).

2 Rooted cuttings being grown as standards should be kept on the move with adequate warmth and water. Do not stop them, and make sure each is supported with a cane. Continue to stop those intended for bush plants when they have made three or four pairs of leaves.

3 Take cuttings from old plants as soon as the shoots have made three or four pairs of leaves. These will make good bush plants for the summer, or you can use some to make a start on trained plants.

April

1 Continue to take cuttings from old plants as soon as the new shoots have made three or four pairs of leaves.

2 Young single-stemmed plants that have already been stopped once will soon produce sideshoots. Stop these at two or three pairs of leaves to produce bushy plants.

3 Make up hanging baskets for the greenhouse.

4 Towards the end of the month, indoor plants that have been stored can be started into growth with more light, water and warmth. Cut back as soon as growth starts and repot a few days later.

5 Remove the protective soil coverings from hedges and 'near-hardies', unless the weather is severe.

6 Pot-grown plants that have been 'trenched' for the winter may be lifted now, after the soil over them has been carefully removed. If they have started into growth, cut them back and either replace

them in the trench or put them indoors or in the greenhouse. If returned to the trench, re-cover with soil almost to their tops.

7 On young standards remove the sideshoots from the main stem until the required height is reached. On older standards the bushy 'head' should be pruned in the same way as bush plants.

8 On all plants keep pests at bay with the help of sprays, fumigants and systemic insecticides.

May

1 Take cuttings of indoor plants as soon as the new shoots have made three or four pairs of leaves. Stop the remainder at the same length to produce new bushy plants.

2 Young plants in 3in pots should be potted on into 5 or 6in ones as soon as the roots show in the drainage holes.

3 Prune trained plants as necessary to keep them in shape.

4 Prepare composts for window-boxes, hanging baskets and other containers, ready for planting up next month.

5 Level and rake the sites for new outdoor plants, ready for use next month, and work in a light dressing of a general fertiliser.

6 Plant up hanging baskets for outside. If the weather is bad, keep them temporarily in a window indoors.

7 Greenhouse plants intended for planting out in the garden or in window-boxes and other containers must be hardened-off first. Stand them in a deep cold-frame and gradually increase the ventilation.

June

1 Trim hedges and the 'near-hardies' by removing all dead wood; then on the latter shorten any remaining stems back to firm wood and cut all the sideshoots from the main branches back to two buds.

2 Plant out new hardy varieties. Also plant out the fuchsias for summer bedding after first hardening them off. Short or imperfect standards can be used as 'dot plants' among the latter.

3 Plant up window boxes, troughs and other containers, and place hanging baskets outside.

4 Continue to take cuttings from both indoor and greenhouse plants. Pot those up that are already rooted.

5 See that indoor plants get adequate air and water, with shade from full sun. Feed sparingly when roots seen in the drainage holes.

6 In the greenhouse, see that plants intended for exhibition are each given ample room. Remove faulty leaves or damaged shoots as soon as you see them.

7 Apply shading to the greenhouse and damp down in hot weather. Plenty of ventilation will also help to keep the temperature down.

8 Continue to keep down pests.

July

1 Take cuttings for growing on as standards and other trained plants. Grow these on without stopping, except in the case of bush plants.

2 On all plants remove faded flowers straight away. On bush plants remove thin, overcrowded shoots.

3 Check that the ties are not 'strangling' trained plants. Renew them where necessary.

4 Place indoor plants outside in mild wet weather to freshen them up, but protect them from the wind.

5 Make sure that plants in hanging baskets and other containers are given plenty of water. If hanging baskets dry out, soak them in a bucket of water. Apply a weak liquid feed about once a fortnight.

6 Trim hedges lightly by shortening all wandering shoots.

August

1 Overhaul all exhibition plants ready for the show date.

2 Visit shows and note new varieties. Collect catalogues for future reference.

3 If cascade varieties in hanging baskets tend to produce their shoots on top of each other, spread them out over a 'fence' of canes and string to provide a more attractive effect (also window boxes).

4 Where greenhouse plants are temporarily placed indoors replace them with others occasionally, to give the original ones a rest.

5 Continue to take cuttings but remember that they will need a temperature of about 50°F (10°C) in winter.

September

1 Start to dry off greenhouse and indoor plants by laying them on their sides outside. Some of the later-flowering ones can be kept going to prolong the season.

2 Trained plants growing permanently in the greenhouse should be

137

given gradually less water, until the leaves fall.

3 At the end of the month remove plants from all types of container, and pot them up ready for drying off. Keep the old compost on them intact and add just enough new compost to fill the pots.

4 See that all plants are labelled before you dry them off, so that you know which is which next year. All rooted cuttings must also be given reasonably permanent labels.

October

1 When all the leaves have fallen, store the dried-off plants in a frostproof place and keep them just moist.

2 Greenhouse, bedding and indoor fuchsias can all be stored by 'trenching'. Dig a trench in a well-drained spot so that it is deep enough almost to cover the tallest plant. Then stand the plants, in their pots, in the trench and cover them with leaves, bracken or peat, covered in turn with a mound of soil along the whole length of the trench. As long as the pots are at least 5in below the surface they will come through most winters safely.

November

1 Give young fuchsia hedges and both new and old 'near-hardies' their winter protection by covering the roots and the base of the plants with bracken, leafmould, peat or ashes. Do not cut back the old stems beyond a light trimming to keep them tidy—they will provide a little extra protection.

2 See that all pot-grown plants are safely stored away on their sides and away from frost.

3 Prepare sites for new 'near-hardies' by digging the ground thoroughly and working in plenty of humus such as leafmould, spent hops or peat. There is no need to add fertilisers yet.

December

1 Cuttings that were rooted and potted up indoors in the autumn must be kept growing through the winter. Stand them on a warm window ledge and keep them just moist.

2 Continue to check stored plants for dryness. If the soil feels dry soak the plant thoroughly from underneath; then leave it until it is almost dry again.

The Varieties:
A Descriptive List

'Abbé Farges' Tube and sepals pale cerise, corolla lilac; an old but very lovely variety with small semi-double flowers with reflexing sepals. Its upright habit and medium growth, plus its exceptional profusion of bloom, make it a fine pot-plant for the greenhouse provided that it is carefully handled to avoid breaking the rather brittle branches.

'Aintree' Tube and sepals pink with a touch of creamy white, corolla a luminous deep pink. A good plant for a basket. Flowers single.

'Alameda' A good double variety for exhibition. The tube and broad upturned sepals are deep crimson and the wide-spreading corolla is white, veined crimson. Makes a strong upright bush and can be also trained as a standard.

'Alaska' Practically a white self, with a faint tinge of pink. A good exhibition variety with large double flowers fairly freely produced, the very full corolla making them among the largest of the white selfs.

'Alice Ashton' Tube and sepals pink, corolla porcelain blue. A small-leaved double-flowered variety with delicate growth of cascade form, making a fine basket plant.

'Alison Ryle' A seedling from the popular Tennessee Waltz, which it resembles except that the corolla is a deeper shade and the sepals are more flaring. A semi-double that can be grown in the same ways as the parent and is probably just as hardy.

'Ambassador' Tube and sepals white, flecked bright red, corolla violet-purple. A fine single that in the greenhouse will make a vigorous bushy plant with large blooms.

'**Amelie Aubin**' Tube and sepals a creamy waxen-white, corolla rose-cerise. Plenty of large single flowers on a strong upright bush suitable for training as a standard or pyramid.

'**American Flaming Glory**' Tube and sepals pink, corolla purple and red, edged with pink and orange. A very showy variety with plenty of large flowers on an upright plant with pale green foliage.

'**Andrew Ryle**' A recent introduction, with small flowers on a shapely bush. Single flowers, tube and sepals red, corolla white.

'**Angela Rippon**' A recent introduction with single flowers that tend to be on the small side. The tube and sepals are greenish-white and the corolla a bluish-lavender fading to purple. Plenty of bloom on a short-jointed bush.

'**Angel's Flight**' Tube white with deep pink sepals, corolla also white, with deep pink stamens. A very attractive double variety; both the tube and sepals are extra long. Makes a good basket plant.

'**Annabel**' Another recent introduction, forming an upright bush with large double flowers. The tube and sepals are white, flushed pink, and the corolla is a rich creamy-white.

'**Ann's Beauty**' A trailer that makes a good basket. Showy double flowers, neyron-rose on the tube and sepals and violet and rose on the large corolla.

'**Applause**' Another recent introduction that makes a good basket. The tube and sepals are cream to pale carmine and the huge corolla is coral-orange.

'**Aquarius**' Tube and sepals pink, corolla bell-shaped and pink, deepening towards the centre. Medium-sized single flowers on an upright bushy plant that needs shade in the greenhouse.

'**Archie Owen**' A good trailer for a basket, with double flowers and dark green foliage. Almost a pink self, with the corolla a softer colour than the tube and sepals.

'**Auntie Maggie**' The tube and sepals are pink, with the underside

140

of the sepals a deeper pink. A fairly recent introduction that makes a good bush or standard.

'**Autumnale**' A cascade variety with plenty of medium-sized flowers, red on the tube and sepals and purple on the corolla.

'**Belle of Salem**' The tube and sepals are deep rose-pink and the saucer-shaped corolla is lavender with pink stripes. Of cascade habit it makes a beautiful basket with plenty of single flowers.

'**Bernadette**' Tube and sepals deep rose-pink, corolla powder-blue. The colouring of the double blooms makes it a well worthwhile variety for the greenhouse, where it will make an upright branching plant with plenty of medium-sized flowers.

'**Blue Pearl**' Tube and sepals white tinged pink, with green tips to the sepals, corolla deep violet-blue. A good double-flowered variety for the greenhouse, of vigorous upright growth.

'**Blue Pinwheel**' The red tube and reflexing sepals and the orchid-blue corolla of the medium-sized single flowers make this an outstanding plant, with a trailing habit that makes it ideal for a basket.

'**Bon Bon**' Tube long and white, terminating in short, broad, pink sepals; corolla pale pink. The medium-sized double flowers are set against glossy foliage and the self-branching trailing habit makes this variety well worth trying as a weeping standard. Also makes a good basket plant.

'**Caesar**' Tube and sepals red, corolla purple fading to burgundy. The very large double flowers are carried on a strong upright plant that can be readily adapted to a basket.

'**Cameron Ryle**' A recent introduction of trailing habit, making it suitable for a basket. The tube and sepals of the frosty-white semi-double flowers contrast with the dark bluish-purple corolla.

'**Capitola**' A fairly modern introduction making a fine trailer for a basket. Its unusual colouring, with a pink tube terminating in green-tipped white sepals and an orange-rose corolla, make it well worth growing. Double flowers.

'**Capri**' Not an easy variety but a real beauty when well grown. The thick, short tube and broad granular sepals are white and the heavy corolla is a deep rich violet-blue. The growth is normally upright but the weight of the huge double flowers often turns the whole plant into a trailer suitable for a basket.

'**Celia Smedley**' A single-flowered variety that makes a fine upright bush, its pale pink tube and sepals making an effective contrast to the bright vermilion corolla. Plenty of medium-sized flowers.

'**Cheers**' A very recent introduction with double flowers on a semi-trailing plant. The flowers are large, with the tube and sepals orange-red and the corolla coral-red.

'**Coachman**' A beautiful single variety with the tube and sepals a pale salmon and the corolla orange-vermilion. The long flowers are really showy and are produced over a long period on a plant that grows naturally into bush form. Well worth growing.

'**Constellation**' A creamy-white self with double flowers which, like all white varieties, are easily marked by water. Apart from this it makes a superb plant with large flowers freely produced on a vigorous upright plant; makes a good standard or half-standard.

'**Curtain Call**' A fine double with the tube and sepals carmine and the corolla rose-bengal, flecked lake and crimson. An unusual feature of the large plentiful flowers is that the petals of the corolla are serrated, while another characteristic is that four flowers are produced from each of the leaf axils instead of the normal two. A good exhibition variety.

'**Daisy Bell**' A recent introduction of unusual colouring, with the tube and sepals of the small, single flowers a pale orange and the corolla vermilion shading to pale orange. The growth is strong and upright and flowering is prolific.

'**Danish Pastry**' A good single variety for a basket. The tube and sepals are coral-pink with the latter tipped green, and the corolla is lavender shading to salmon-pink.

'**Dark Eyes**' An erect bushy variety with double flowers. The short

142

tube and broad upswept sepals are a deep red and the corolla is a deep violet-blue, with the petals rolled and curled. Free-flowering.

'Dark Secret' The double flowers are beautifully coloured, with the tube and sepals a greenish waxy-white and the corolla a rich violet-blue with phlox-pink petaloids surrounding it. A fine exhibition variety of upright habit with plenty of bloom.

'Dee Dee' Another fairly recent introduction, with medium-sized double flowers on a plant of cascade habit suitable for a basket. The tube and sepals are white, with the latter long and curved, tipped green, and the corolla is rose-purple.

'Derby Imp' A fine single of self-branching habit that makes a good basket or bush. A profuse flowerer, with the tube and sepals crimson and the corolla violet-blue.

'Dilly Dilly' A double of semi-trailing habit with large flowers freely produced. The lilac-blue corolla shows up well against the pink tube and sepals.

'Display' A very old single that can be successfully grown indoors, although it is also suitable for most forms of training. Almost a pink self, with the deeper pink corolla somewhat saucer-shaped. Plenty of bloom on an upright bushy plant. Also successful as a summer bedding plant.

'Drame' An old and hardy variety with single flowers, scarlet on the tube and sepals and violet-purple on the corolla. Foliage yellowish-green. Claimed to be hardy almost everywhere and may be used as a hedge in mild districts.

'Dr Foster' One of the largest-flowered of the hardy varieties, this was introduced about the turn of the century. Of a rather spreading habit it makes a colourful bush to about 3ft high, with the tube and sepals of the single flowers scarlet and the corolla purple, veined scarlet.

'Dr Jill' A quite recent introduction that makes a good upright bush with double flowers, pale purple-red on the tube and sepals and a delicate pale lavender-pink, with deeper veinings, on the corolla.

Dunrobin Bedder' This hardy variety is particularly useful for its dwarf, spreading habit which makes it suitable for the rock garden or for path edging; or it will make a good occupant of a window-box. The single flowers, with the tube and sepals scarlet and the corolla deep purple, are small but prolific. Not easily available.

'Dutch Mill' The shape of this flower makes it most attractive, with the long rose-pink tube and sepals and the equally long bell-shaped corolla of pale violet, shading towards the base. The plant itself makes a very free-flowering medium-sized bush.

'Easter Bonnet' A large-flowered double on which the tube and sepals are pink, with the latter tipped green, and the corolla is a dusky rose-pink, slightly darker at the edges. Plenty of bloom on a bush of medium size.

'El Camino' A fine upright grower that makes a good standard. The recurving sepals of the fairly large double flowers are broad and short, and of the same rose-red as the tube, while the corolla is white, flecked and veined rose. The flower is also made more attractive by the shape of the corolla, which consists of large central petals surrounded by smaller and more spreading ones.

'Empress of Prussia' A first-class hardy variety, going back to mid-Victorian times. Growth is short, strong and erect, up to about 4ft, and for a hardy variety the flowers are quite large. The tube and sepals of the single flowers are vivid scarlet and the corolla is scarlet lake.

'Eternal Flame' A very free-flowering single, with the tube and sepals salmon-orange and the corolla a smoky-rose streaked at the base with salmon. Makes a good upright bush.

'Ethel' An attractive large-flowered double with the tube and sepals white, the latter pink on the undersides, and the corolla lavender with surrounding pink petaloids. A good upright bush.

'Eva Boerg' Forms a very free-flowering low bush with medium-sized flowers varying from single to semi-double. The tube and sepals are white, faintly tinged pink, and the corolla is violet-purple. Makes a good basket plant.

'Evensong'   An excellent white self, with the sepals reflexing to hide the tube. Makes a fine free-flowering upright bush.

'Fan Dancer'   A striking variety, with double flowers; tube and sepals are red and the corolla orchid-blue splashed with pink and red. The growth is vigorous and upright and the flowers are somewhat rectangular in shape.

'Fiona'   An excellent basket variety with the tube and long graceful sepals of the single flowers white and the long corolla blue, aging to purple. An upright and densely bushy variety with plenty of bloom.

'Flamboyant'   An aptly named cascade variety that makes a fine basket. The tube of the semi-double flowers is red and the sepals white, while the flaring corolla is cerise and purple. The large flowers are freely produced.

'Flaming Glory'   Another fine basket variety, of trailing habit. The tube and sepals are pink and the centre of the corolla is purple and bright red, while the outer edges of the petals are pink and orange. The flowers are double and the foliage is a pleasing light green.

'Flash'   A fairly hardy variety suitable for a window box. A red self, with small flowers very freely produced on a small self-branching bush. The habit is low and spreading.

'Flavia'   The tube and sepals of the large double flowers are a rosy-pink and the corolla a deep lilac. An excellent basket variety, with plenty of bloom.

'Flirtation Waltz'   A vigorous upright variety that makes a beautiful bush with plenty of double flowers of good size and substance. The tube and sepals are creamy-white, flushed pink, and the corolla is shell-pink.

'Fluffy Frills'   A pink self, with the corolla ruffled and frilly. Its double flowers and semi-trailing habit make it a good basket variety.

'Formossima'   A very old variety that is still well worth growing, if only for its prolific flowering. The long single flowers, white with a tinge of pink on the tube and sepals, and rose-pink on the corolla, are very lovely, and the plant itself is vigorous and upright, making a handsome standard.

'Forward Look' An attractive variety that makes a fine free-flowering bush. The light pink tube terminates in china-rose sepals, tipped green, and the corolla is a charming shade of wisteria-blue, shading to violet.

'Foxtrot' A semi-double with the tube and sepals a pale cerise, tipped green, and the corolla pale lavender, pink at the base. Makes a bushy plant with medium-sized flowers.

'Frosted Amethyst' A trailing variety which makes a good basket. The tube and sepals of the double flowers are red and the corolla amethyst-purple, flecked with pink.

'Gay Anne' A self-branching bush or trailer. The semi-double flowers, claret-rose on the tube and sepals and mallow-pink with a magenta-rose picotee edge on the corolla, are medium-sized and freely produced.

'Gay Fandango' A popular variety that makes a good bush or standard. The tube and long, wide-spread sepals are rosy-carmine and the long many-petalled corolla is rosy-magenta. The large double flowers are produced freely on a plant that is upright and bushy.

'Gay Parasol' A very recent introduction with the tube and sepals of the double flowers ivory-pink and the corolla deep purple fading to burgundy. The plant can be trained into either an upright or a lax bush, and could well be tried as an espalier.

'Gay Paree' A large-flowered double with the tube and sepals white, flushed carmine and tipped with green, and the corolla a mixture of purple, pink and carmine. Plenty of bloom on a low, lax plant.

'Gay Senorita' A useful variety for early flowering. The tube and sepals are rose-red and the bell-shaped corolla is a dark lilac rose. Makes an upright bushy plant with plenty of medium-sized flowers.

'General Wavell' A strong and free-flowering bush with medium-sized double blooms. The tube and sepals are pale cerise and the very long corolla is salmon shading to magenta. The upright habit also makes it suitable for growing as a half-standard or standard.

'Genii' (syn 'Jeane')   The single flowers are pale cerise on the tube and sepals and scarlet on the corolla, and while they are small there are so many of them that they make a wonderful display. Some catalogues claim this to be a completely hardy variety and it is certainly worth trying outside for its profuse flowering and attractive yellowish-green foliage.

'Georgia Peach'   A recent introduction whose semi-trailing habit makes it a fine basket plant. The tube and sepals of the semi-double flowers are white and the corolla is peach-pink, marbled white.

'Glenn-Mary'   A trailing variety with double flowers on which the tube and sepals are pink and the corolla rosy-violet. The large blooms are freely produced.

'Glitters'   An easy single variety of attractive colouring. The tube and sepals are waxy white except for the salmon underside of the latter and the wide-spread corolla is a vivid orange-red. Makes a strong tall plant with a profuse display of medium sized flowers.

'Golden Dawn'   A lovely variety with plenty of medium-sized single blooms on an upright bushy plant. The tube and sepals are a light salmon and the corolla is a pale orange aging to pink.

'Golondrina'   A fine variety with the tube and sepals rose-madder and the corolla magenta, edged rose. The shapely buds open to large single flowers, with long reflexing sepals produced freely on a plant that makes a fine basket or standard.

'Gracilis variegata'   A hardy variety with the tube and sepals of the single flowers bright red and the corolla purple. The foliage of this strong upright plant is variegated silver and pink, making it most attractive.

'Groovy'   An upright or trailing variety with bell-shaped blooms, the tube and sepals coral and the corolla fuchsia-pink to magenta. Medium-sized single blooms.

147

'Guinevere' An easy and popular variety for both exhibition and greenhouse culture. The large semi-double flowers are produced on a strong upright bush that is easily kept to a good shape and which can also be trained as a standard. The white tube and sepals and the orchid-blue corolla make an effective contrast.

'Gus Niederholzer' A striking fuchsia with very large double flowers fairly freely produced. The tube and sepals are carmine, with the latter long and upswept, and the corolla is a scalloped bell-shape of veronica-blue, aging to orchid-pink.

'Happy Fellow' Makes a strong upright bush with the medium-sized flowers produced in clusters. The tube and sepals of the single flowers are a light, clear orange and the corolla is smoky-rose.

'Hawaiian Night' An upright or trailing variety. The tube and sepals of the double flowers are waxy-white and the large corolla a deep orchid-mauve. Grown as a trailer it makes a good basket or window-box plant.

'Hawkshead' This single variety is said to be an improved version of the very hardy *magellanica* 'Alba', but whether its greenish-white flowers are an improvement is a matter of taste.

'Healdsburgh' A bush or semi-trailing variety with large double blooms, the tube and sepals light red and the corolla light purple.

'Heart Throb' A useful upright variety with double flowers on which the tube and broad sepals are white, with a pink tinge on the underside of the latter; the corolla is an attractive medium blue, with large and small petals curled and folded. The whole forms a wide-spread, almost flat flower, white at the centre.

'Heidi Ann' The tube and sepals on this useful variety are cerise while the corolla is pale orchid-purple, with the flowers showing up well against the dark green foliage. The plant makes a strong upright bush covered with the double flowers.

'Heirloom' A bush or trailer, with large double blooms, pink on the tube and sepals and with the corolla marbled blue and pink. An attractive variety that can also be used in a basket.

148

'Heritage' An old semi-double of upright habit that claims to be hardy, at least in mild areas. The tube and sepals are scarlet and the corolla rich purple, and the plant itself is well worth trying both under glass and outside.

'Heron' A good variety to grow as a standard under glass, although it also makes a fine summer bedding plant, particularly as the flowers are weather-resistant. Makes a strong bushy plant on which the tube and sepals of the large single blooms are scarlet and the corolla scarlet, veined cerise.

'H. G. Brown' The flowers are small but plentiful on this quite hardy variety. The tube and sepals of the single flowers are a deep scarlet and the corolla Indian lake, while the plant itself is low and bushy, with dark green foliage.

'Hidcote Beauty' A fuchsia of unusual colouring, with the tube and sepals of the single flowers a waxy-cream flushed and tipped with green and the corolla a clear salmon-pink. Forms an upright bushy plant with pale green foliage against which the medium-sized flowers show up well.

'Hi-Jinks' A low-growing double-flowered variety with the tube and sepals white, with a pale pink reverse on the latter. The corolla is dianthus-purple, marbled white. The low, somewhat trailing growth makes it suitable for a basket.

'Hollywood Park' This American variety is said to be hardy, although it is certainly not one of the better-known hardy varieties. The medium-sized semi-double flowers, with the tube and sepals cerise and the corolla white, veined pink, are produced very freely on an upright bushy plant.

'Hombre' A fine trailer with large double flowers. The tube and sepals are bright pink and the corolla lavender with splashes of pink. Makes a good basket.

'Hula Girl' A pretty trailer with double flowers on which the tube and sepals are rose and the corolla forms an attractive white 'hula skirt'. Medium-sized flowers.

'Icicle'   A white self except for the pink stamens of the corolla. The flowers are double, and the trailing habit, combined with the rather small light green foliage, makes this an excellent variety for a basket.

'Igloo'   A large-flowered double with the tube and sepals pink and the corolla white, with flaring petals. A good trailer for a basket.

'Igloo Maid'   A white self. The large fully double flowers are produced on a strong upright plant with attractive foliage of a light apple-green.

'Impudence'   An attractive variety on which the four petals of the corolla open out almost flat. The tube and sepals of the single flowers are red and the corolla white with pink veining. An upright grower with plenty of bloom.

'Inspiration'   A fairly easy variety, and a good one to start with for exhibition. The vigorous upright growth and profuse flowering make it an outstanding plant, with the medium-sized double flowers almost a pink self, except that the corolla is slightly darker than the tube and sepals.

'Interlude'   A superb basket variety with a lavish display of medium-sized double flowers. The slender tube and the sepals are white, flushed pink, and the multi-coloured corolla is a blend of pink, purple and red.

'Italiano'   A very free-flowering double with medium-sized blooms on a plant of trailing habit. The flowers are beautiful, with the tube and sepals salmon-pink and the very full corolla a rosy-purple fading to burgundy.

'James Travis'   A double-flowered variety that is claimed to be hardy. The bright red of the tube and sepals and clear blue of the corolla make it unusually attractive for a hardy variety. The fairly large flowers are freely produced on a tall, spreading plant.

'Jandel'   A recent double introduction with a semi-trailing habit, suitable for a basket. Tube and sepals pinkish-white, corolla orchid-blue.

'Jean Burton'   A fine single that makes a strong sturdy bush. The large flowers are bell-shaped, with the tube pale pink and the backward-turning sepals rhodomanthine-pink. Corolla pearl-white. An excellent greenhouse variety but does best in the shade.

'Jeane'   See 'Genii'.

'Jeanette Broadhurst'   A beautiful variety for a basket. The tube and sepals are rose-madder and the corolla is marbled rose-madder and mauve. The growth is cascade, with a prolific show of single flowers.

'Joanne'   A recent introduction of trailing, self branching habit. The tube and sepals of the single flowers are dusty-pink and the corolla mauve. Free-flowering.

'Joan Pacey'   A free-flowering self-branching bush. The tube is white and the pink sepals are tipped with green, while the corolla is phlox-pink.

'Jose Joan'   A modern variety with the tube and sepals white and the corolla pale violet. The double flowers are large but the plant needs a lot of pinching to train it to shape.

'Joy Patmore'   A carmine-and-white single that makes a good bush or standard. Said to be hardy.

'Jubie-Lin'   A cascade variety with huge double flowers freely produced. The tube and sepals are red and the corolla very dark purple.

'Jupiter '70'   An outstanding single variety with the tube, sepals and corolla all in different shades of pink. Small flowers plentifully produced on a strong upright bush.

'Kaleidoscope'   A large-flowered double that lives up to its name in that the very large corolla is in various shades of purple, red, pink and lavender. The tube and sepals are red. An upright grower.

'Kathleen'   A double variety that makes a fine bush or standard. The tube and sepals are red and the corolla orange-white. Self-branching.

'Kathy Louise'   A vigorous trailer with dark green foliage and non-fading double flowers. The tube and sepals are carmine and the corolla soft rose.

'Kernan Robson'   A fine exhibition variety, of a colouring that makes it outstanding. The tube and sepals are flesh-pink, the latter red on the inside, and the beautiful corolla is large and fluffy and a distinctive smoky-red. Plenty of large flowers on an upright bush that should make a good standard.

'Kings Ransom'   An excellent variety with large double flowers freely produced on a strong upright bush. Can also be grown as a standard and makes a fine exhibition plant. The tube and sepals are an almost transparent white and the corolla a rich dark purple.

'Kiwi'   A semi-double with the tube and sepals white and the corolla china-rose with pale pink marbling. A trailer suitable for a basket.

'La Bianca'   Another white self, except for a faint tinge of pink on the corolla. Makes a profuse show of medium-sized flowers on a plant of bushy upright habit.

'Lady Anne'   The fairly large flowers are produced very freely on a strong upright bush. The tube and sepals are white, tipped green, and the corolla purplish-blue marbled with blue and pink.

'Lady Isobel Barnet'   A most attractive variety of exceptionally free-flowering habit, with at least eight blooms to each leaf axil. The tube and sepals are rose-red and the corolla is rose-purple flushed at the edges of the petals with imperial purple.

'La Fiesta'   A good plant except for its lax growth, which needs a lot of training to keep it in shape, although it can be more easily trained as a basket plant. The tube and sepals of the double flowers are white, flushed with pink, and the corolla is a blend of white, red and purple.

'La France'   A very old variety that still remains popular. The tube and sepals of the large double flowers are a rich scarlet and the very full corolla is a rich violet-purple. Makes a strong upright bush.

'La Neige'   Almost a white self except for a touch of pink on the corolla of the double flowers. Of lax growth it makes a fine basket with plenty of medium-sized blooms. The white flowers make an attractive change from the colours usually associated with basket varieties.

'L'Arlesienne'   A very attractive flower with the pale pink sepals reflexing and the long, even paler petals of the corolla taking on a beautiful shape. Semi-double.

'La Rosita'   Makes a good upright grower with plenty of medium-sized double flowers, with the tube and recurving sepals rose-pink and the corolla orchid-pink. Makes a good standard.

'Laura'   The ivory-white of the tube merges into the neyron-rose of the sepals, while the corolla is fuchsia-pink, with the petals edged red. The flowers are large and shapely. Makes a neat and easy bush.

'Laurie'   A fine trailing variety for a basket, with double flowers on a vigorous bush plant. The tube and sepals are a pale pink and the corolla is a pale orchid-pink, with spreading petaloids.

'Lavender Lady'   Apparently named from the purple-violet corolla of the single flowers, although the tube and sepals are bright red. Makes a good upright bush suitable for training as a pillar or pyramid.

'Lena Dalton'   A double that makes a good upright and rather small bush with plenty of medium-sized blooms. The tube and recurving sepals are pink and blend with the pink-flushed blue corolla.

'L'enfante Prodigue'   An old hardy variety sometimes listed as 'Prodigy', that will grow almost anywhere. Makes an upright bushy shrub with plenty of medium-sized flowers, cerise on the tube and sepals and royal purple on the corolla.

'Leonora'   A very soft pink self with single blooms in profusion on an upright bush. The corolla is attractively bell-shaped.

'Libra'   A good trailer for a basket, with large double flowers freely produced. The tube and sepals are pink and the corolla varies from pale blue to lavender, with splashes of pink.

'Lilac Lustre' An upright self-branching bush with double flowers, rose-red on the tube and sepals and powder-blue shading to lilac on the corolla. The rich green foliage adds to the appearance of this variety.

'Lilac 'n' Rose' A lax-growing plant that is more suitable for a basket or as an espalier. The long double flowers are a blend of various shades, with the rose-pink tube and sepals and the pale lilac corolla taking on a deeper colour as they age.

'Lilibet' Its cascade habit of growth makes this an excellent basket plant, with plenty of long, pointed buds opening to attractive double flowers. The long tube and the recurved sepals are white, flushed carmine, and the corolla is a soft rose, shaded crimson lake.

'Linda Copley' A clear pink self that makes a strong upright bush with single flowers. Ideal for the greenhouse.

'Lindisfarne' A recent introduction with semi-double flowers on which the tube and sepals are pink and the corolla is a rich dark violet shaded pale pink. Makes a vigorous upright bush of self-branching habit.

'Lisa' A good double-flowered trailing variety for a basket. The tube and sepals are rose and the corolla lavender-blue.

'Little Beauty' An upright grower with small single flowers that are pleasing enough to make this a good exhibition variety. The tube and sepals are rose-pink and the corolla lavender.

'Little Ronnie' A modern semi-double variety with plenty of bloom over a long season. Makes a good pot-plant or half-standard. The tube and sepals are rose and the small corolla a deep blue.

'Lolita' The tube and sepals are a pinkish-white and the corolla an attractive porcelain-blue on this excellent trailer. The double flowers are large and freely produced.

'Lollypop' A free-flowering variety with the tube and curving sepals of the single flowers a shining pink and the corolla a deep pink aging to paeony-purple. Owing to its lax habit it is most successful in a hanging basket.

154

'**Lord Lonsdale**' Makes a good bush variety with medium-sized single flowers freely produced. The pale apricot tube and sepals and the deep salmon-orange corolla make this a very pleasing variety, all the more so because of the unusual curling and crinkling of the foliage.

'**Lord Roberts**' A good exhibition variety of strong upright growth. The flowers are single, with a very large corolla of deep purple which contrasts well with the scarlet tube and sepals. Early flowering.

'**Lothario**' The large double flowers are pink on the tube and sepals and marbled purple, red and pink on the corolla. A good basket variety of vigorous trailing habit.

'**Loveliness**' A bushy upright grower that makes a fine standard. The profusion of single flowers, ivory-white on the tube and sepals, with the corolla cerise-carmine, makes this a delightful variety, with the extra advantage that it is quite easy to grow.

'**Lucky Strike**' A fine semi-double that grows quickly into a strong upright bush. It also makes a good standard, with plenty of large flowers of attractive colouring, pale pink on the tube and sepals, and blue marbled with pink on the corolla.

'**Magenta Flush**' A good upright grower that should be tried as a standard. The tube and sepals of the large double flowers are red tipped with green and the corolla is in two shades of red.

'**Maharajah**' A large-flowered double of upright growth. The salmon-pink tube and sepals contrast well with the dark purple corolla, marbled salmon and orange.

'**Major Heaphy**' Forms a low bush with small flowers in profusion. The tube and sepals of the single blooms are brick-red and the corolla scarlet. Of little use as an indoor plant as it will not stand a dry atmosphere.

'**Mandarin**' The large flowers are produced very freely on this semi-double variety which makes an upright plant in the open but a trailer under glass. The flesh-pink tube and sepals and the orange and carmine corolla show up well against the leathery dark green foliage.

155

'Margaret'   A semi-double hardy variety with plenty of medium-sized blooms. Makes an attractive plant with the tube and sepals scarlet and the corolla violet, with white at the base of the petals and red veining. Has succeeded as a hedge in some areas.

'Margaret Brown'   Like 'Margaret' this has succeeded as a hedge in some areas. The colouring, with the tube and sepals rose-pink, and the corolla a pale rose-bengal, is unusual for a hardy variety, and although the flowers are small they are produced liberally on a strong-growing bush.

'Margaret Roe'   A hardy variety that makes an attractive bush with single flowers produced freely and continuously. The tube and sepals of the medium-sized blooms are rose-red and the corolla pale purple.

'Margie'   An unusual and striking double flower, with the tube and sepals pale pink and the pale lavender corolla decoratively ruffled and serrated. Makes a strong upright plant with the large blooms freely produced on willowy branches.

'Marietta'   Forms a low bush with large flowers in plenty. The tube and sepals of the double blooms are bright carmine and the corolla is a deep magenta, splashed with carmine.

'Martin's Delight'   A strong upright grower with double flowers, red on the tube and sepals, with the corolla a mixture of red and purple.

'Martin's Midnight'   An exhibition variety of beautiful colouring, with the tube and sepals of the double flowers a bright red and corolla midnight-blue. Makes an upright free-flowering bush of strong growth.

'Martyn Smedley'   The single flowers of this upright bush variety are a waxy rose-pink on the tube and sepals, while the corolla is kingfisher-blue on the inside and wisteria-blue, shaded white, on the outside. Unlike some varieties it tolerates full sun.

'Mary Ellen'   A good trailer for a basket, with large double flowers in profusion. The tube and sepals are pale pink and the corolla lavender-blue.

The profuse, red and white flowers of the almost-hardy 'Snowcap', which is successful as a pyramid

(*right*) A fine pyramid, the cerise and purple 'Brutus'

(*left*) The waxy-white tube and flared carmine corolla of 'Joy Patmore', a bush or standard

Another splendid, though different, pyramid – the old variety 'Molesworth', which has scarlet-cerise and white flowers

'**Masquerade**' A cascade type with large flowers, with the tube and sepals pink and the corolla medium-purple marbled with pink, the outer petals pleated and opening wide as they age.

'**Mauve Beauty**' A very old variety but still one of the best in its colour range. The tube and sepals of the double flowers are a bright glowing red and the corolla is lilac-mauve, slightly veined red. A strong free-branching grower, with medium-sized flowers in profusion. Its low growth makes it ideal for window-boxes.

'**Mauve Poincaré**' The tube and sepals are crimson and the corolla is pure mauve on this excellent single variety, which blooms early and continues for a long time. The many flowers are large and of attractive shape, with the corolla long and bell-like.

'**Meadowlark**' A modern variety with semi-double flowers, white on the tube and sepals, and with a ruffled corolla shading from lavender to rose. The medium-sized flowers are produced on a medium-sized bush.

'**Mendocino Rose**' A very large-flowered variety forming a vigorous upright bush. This is an excellent recent introduction with the tube and sepals of the double flowers coral-pink and the corolla rose and coral.

'**Mephisto**' A single-flowered variety for exhibition. The small flowers, with the tube and sepals red and the corolla a deep crimson, make up in number what they lack in size and the plant itself soon makes a large specimen.

'**Mieke Meursing**' Makes a very early and free-flowering bush of compact habit. The tube and sepals are red and the corolla an unusual shade of pink. The long stamens add to the appearance of the single blooms.

'**Miss Great Britain**' A free-flowering but late single, with medium to large blooms. The pink tube merges into creamy-white sepals with green tips and the corolla is wisteria-blue shading into a deeper blue. Makes a good upright bush.

'**Miss Leucadia**' A large-flowered trailer with the whole flower a

161

soft pink, picotee-edged on the corolla. The slightly deeper shade of the corolla enlivens the colouring. The double flowers are carried on a trailing plant, ideal for a hanging basket.

'**Miss Prim**'   A good indoor plant with semi-double flowers on a self-branching bush. The tube and sepals are carmine and the corolla deep purple becoming paler towards the base.

'**Miss Washington**'   The tube and sepals are a deep rose-pink and the corolla sparkling white, veined pink, on this attractive double. Makes a bushy plant with a profusion of large flowers. Trailing habit, ideal for a hanging basket.

'**Molesworth**'   An old variety that makes a strong upright and dense bush with plenty of bloom. The large double flowers are scarlet-cerise on the tube and sepals and the white petals of the very full corolla are veined pink at the base. Can be trained to any shape.

'**Monsieur Thibaud**'   Another old variety that makes a fine bush or standard. A quick grower, it soon forms a large bush with a liberal display of single flowers, cerise on the tube and long recurving sepals, magenta on the corolla.

'**Moorland Beauty**'   Makes a good bush or standard with semi-double flowers. The tube is red and the sepals neyron-rose, while the corolla is violet shading to a reddish-purple.

'**Morning Light**'   A very showy variety for exhibition. The tube and base of the sepals are coral-pink and the wide upcurving sepals are white, flushed pale pink on the underside. The fully double corolla is lavender-blue with the small outer petals splashed with pink. The large double flowers are freely produced on a plant with attractive pale green foliage.

'**Morning Mist**'   The lax growth makes this suitable for an espalier or weeping standard. The profuse single flowers are orange-rose on the long tube and sepals and the corolla is orange-red suffused purple.

'**Mr Big**'   The tube and sepals of the double flowers are light red and the corolla purple, marbled magenta. A trailer with large flowers, it makes an excellent basket.

'Mrs J. D. Fredericks' A good variety for growing as a bush, standard or espalier. The salmon-pink flowers, with a deeper corolla, are small but are carried in clusters that make a fine show. Growth is strong and upright with light green foliage.

'Mrs Lovell Swisher' A well-tried fuchsia that makes a vigorous bush with small single flowers that make up in quantity what they lack in size. The tube and sepals are ivory-white and the corolla shades from pink to rose-red. Makes a good pyramid or standard.

'Mrs Susan Pugh' A recent introduction that makes a strong self-branching bush with single flowers of pleasing colouring. The tube and sepals are orange, flushed cerise, and the corolla is a similar colour overlaid with purple.

'Mrs W. P. Wood' A good summer bedder, with a profuse display of small single flowers on a stiff upright bush. The tube and sepals are pale pink and the corolla pure white.

'My Fair Lady' A good grower making a strong upright bush with double flowers, strawberry-red on the tube and sepals, lavender, aging to red, on the corolla.

'My Honey' Single flowers in profusion on a trailing plant suitable for a basket. The tube and sepals are a light pink and the small corolla is cyclamen-purple variegated with carmine.

'Native Dancer' A showy trailer with large flowers. The tube and sepals are red and the corolla deep purple.

'Nautilus' An old variety with large flowers freely produced on an upright bush. The tube and sepals are cerise and the corolla white, with light cerise veining.

'Navajo' A fine exhibition fuchsia, with large double blooms freely produced. The rich dark green foliage, carried on a strong upright bush, makes a fine setting for the showy colouring of the flowers, pale reddish-orange on the tube, with green-tipped sepals, and a similar but deeper shade on the corolla.

'Nell Gwynn' A recent introduction with single flowers, orange-salmon on the tube and sepals and with the corolla a vivid orange, edged vermilion. An early-flowering variety with medium-sized blooms on a strong upright bush.

'**Newhope**' A beautiful fuchsia with a cascade of large double flowers. A white self enhanced by the pale pink stamens.

'**New Horizon**' Makes an upright bush with plenty of large flowers. The tube and sepals are pale pink and the corolla a very light blue, a colouring that is at its best when the plant is grown in shady conditions.

'**Neve Welt**' A hardy variety with small but brilliant single flowers, the tube and sepals a rich red, the corolla parma-violet.

'**Nicola**' A fine upright and bushy plant with large single flowers freely produced. The tube and sepals are a rich scarlet and the open saucer-shaped corolla is a lovely velvety blue. Can also be grown as a standard.

'**Normandy Bell**' Makes a strong upright bush with single flowers on which the tube and sepals are a pinkish-white and the bell-shaped corolla a light orchid-blue. The plentiful flowers last well and associate well with the light green foliage.

'**Norvell Gillespie**' The tube and sepals are rose-pink and the corolla a dark orchid shading to magenta on this trailer. Makes a fine basket with plenty of bloom. Double.

'**Novella**' Another good trailer for a basket, with medium-sized double flowers freely produced. The tube and sepals are a rosy-pink and the corolla salmon-orange.

'**Old Rose**' Makes a tall upright bush or standard. The tube and sepals of the double flowers are carmine-rose and the corolla neyron-rose. The name suggests both colour and shape.

'**Old Smoky**' The colour of the double blooms is unusual, with the tube and sepals being flesh-pink and the corolla old rose with a smoky tinge. The large flowers are produced fairly freely and the plant makes a free-branching upright bush. A good exhibition variety.

'Orange Crush' An early-flowering variety with single flowers that are almost a self-coloured orange, except that the corolla is a shade darker than the tube and sepals. Makes a good bush.

'Orange Mirage' A single with strong horizontal growth that makes it a good basket or window-box plant. Another almost self-coloured orange with the corolla darker and redder.

'Oriental Sunrise' A recent introduction with the tube and sepals a lighter orange than the corolla. A showy free-flowering single that makes a beautiful basket. Semi-trailer.

'Other Fellow' A good variety for indoors, with the single flowers having the tube and sepals ivory-white and the corolla coral-pink, shading to white at the base. A dainty free-flowering fuchsia that makes a bushy upright plant.

'Pacific Queen' A relatively easy grower with double flowers, tipped white on the wide phlox-pink sepals and with the fully double corolla a warm shade of old rose. Makes a fine upright plant with dark green foliage. A good variety for exhibition.

'Pantaloons' A semi-double with the tube and sepals light red and the corolla plum-burgundy. Medium-sized blooms on a plant that may be grown as either a bush or trailer.

'Papoose' A very useful fuchsia suitable for training to any shape. The flowers, with the tube and sepals bright red and the corolla very deep purple, are quite small but make up for this by their number, produced throughout a long season.

'Party Frock' A good exhibition variety with large flowers freely produced. The rose-red tube and upcurved sepals and the pale lilac corolla, shaded flesh-pink on the outer petals, make an attractive combination against the dark green foliage. Makes a large bush.

'Passing Cloud' An upright grower with plenty of quite large blooms. The single flowers are rose-madder on the tube and sepals and the corolla is amethyst-violet with white markings.

'Patty Evans' A good one for exhibition with the whole bloom a

165

blend of white and pink. The blooms are large and fairly liberally produced and the foliage is an attractive light green. Can also be grown as a standard.

'Peggy Ann'   A recent introduction of cascade habit. The tube and sepals of the double flowers are red and the large corolla is lilac-purple fading to magenta.

'Peppermint Stick'   A double with the tube and upturned sepals carmine, the latter having a distinct white stripe down the centre of each, hence the name. The corolla has centre petals of a rich royal-purple and the outer petals light carmine. The large flowers are produced fairly freely on a strong upright bush and the plant may also be grown as a standard. A good exhibition variety.

'Perry Park'   A hardy variety that makes a small compact bush with plenty of medium-sized flowers. The tube and sepals are pink and the corolla a blend of lilac and purple.

'Pharaoh'   The profuse single flowers are of medium size on an upright self-branching bush suitable for exhibition or for growing as a standard. The tube is rose-bengal and the sepals white, edged rose-bengal and tipped with green. The corolla is plum-purple aging to ruby-red.

'Pink Cloud'   Makes a strong upright bush suitable for most forms of training. Almost a pink self, with wide, upturned sepals curling slightly and a corolla that opens out wide as it ages. One of the largest of the pinks, with attractive long-pointed buds.

'Pink Flamingo'   A lax grower that needs a lot of tying to keep it in shape, although it makes a fine weeping standard. The tube and sepals are deep pink, with the sepals curling in every direction, and the corolla is a very pale pink with darker veining.

'Pink Marshmallow'   An almost self-coloured pink with large double flowers on a plant of trailing habit. The elongated corolla gives the blooms a distinctive appearance.

'Pink Quartette'   A large-flowered variety of strong upright habit, with semi-double blooms almost self-coloured, with the corolla a paler pink than the tube and sepals. The name is taken from the four rolls of petals that make up the corolla. A very showy variety.

'Powder Puff' A lax-growing plant that can be grown either as a basket plant or as an espalier. There are plenty of fairly large double blooms, Tyrian rose on the tube and strong recurving sepals, and a clear pink on the corolla.

'Pretty Grandpa' A large-flowered semi-double of lax habit, suitable for training as either an espalier or basket plant. The tube and sepals are pink tipped with green and the corolla is white, veined pink.

'Purple Heart' A very large-flowered variety but one that can be temperamental, with only a few flowers, not all of which open. Makes a strong upright bush with double flowers, the tube and sepals pale crimson and the corolla rich violet, streaked with pink and red.

'Quasar' A good cascade variety with large blooms. The tube and sepals are white and the corolla violet. A recent introduction.

'Queen of Hearts' A lax grower suitable for training as an espalier or basket plant. The double flowers are very variable in colour, with the tube, sepals and corolla ranging through red, lilac, mauve and pink. Very large blooms.

'Queen Mary' An easy variety whose sturdy upright habit makes it a fine bush or standard. The single flowers are large and predominantly pink, with the corolla a rather deeper pink, shaded mauve.

'Rachel Catherine' A very recent introduction forming a strong self-branching bush with single flowers, the tube and sepals deep red and the corolla rich purple. The prolific flowers are flared and show up well as they stand clear of the foliage. One of the most recent introductions.

'Raggardy Ann' Very aptly named, as the corolla of the double blooms has a very ragged appearance, with a circle of blue petals surrounded by pink ones. The tube and sepals are cerise and the plant makes a good variety for the greenhouse, with plenty of large flowers on a strong upright plant.

'Raintree Legend'   A very free-flowering semi-double of semi-trailing growth; tube and sepals pale pink and corolla creamy-white. A good basket plant.

'Raspberry'   An upright grower with plenty of large double blooms. The two shades of pink, pale on the tube and sepals and much darker on the corolla, make a pleasing combination.

'Red Shadows'   A fine double with the tube and sepals of the very large flowers a rich crimson and the ruffled corolla a deep burgundy-purple, changing with age to a ruby-red. The lax habit makes it suitable for training as an espalier or for a basket.

'Ridestar'   An upright self-branching bush of a floriferous habit that makes it a good exhibition plant. The double flowers are red on the tube and sepals and a deep violet-blue on the corolla.

'Romance'   Of cascade habit this makes a fine basket with plenty of large double flowers on which the tube and sepals are pink and the corolla a pale violet-blue.

'Rosecroft Beauty'   The attractive foliage, sage-green and yellow, makes this a very attractive plant. Of bushy upright habit it produces plenty of rather small semi-double flowers, with the tube and sepals bright red and the corolla white, veined red.

'Royal Touch'   A lax-growing plant readily trained as a basket or espalier. The tube and sepals of the freely produced large double flowers are white with a pink reverse, and the corolla is royal-purple.

'Royal Velvet'   A superb exhibition fuchsia, normally a self-branching bush but also capable of being trained as a basket plant or standard. The large double flowers are very freely produced and each one is a picture in itself, with the tube and upswept sepals being crimson-red and the ruffled corolla opening to a deep purple changing to purplish-crimson, further enhanced by long crimson stamens.

'Ruffles'   A cascade variety with plenty of fairly large double flowers. The tube is deep pink, the sepals the same colour but tipped with green, and the very open ruffled corolla is violet with pink petaloids.

'**Ruthie**'   A very free-flowering upright grower with medium-sized double blooms. The tube and large crinkled sepals are white, flushed red, and the corolla is reddish-purple with pink petaloids.

'**Ruth King**'   A fine trailer with double flowers, the tube and sepals pink and the corolla lilac and white. Makes a good basket.

'**Sally**'   A trailer with a profuse display of medium-sized double flowers, tube white, sepals white with a pale pink reverse, and corolla lavender.

'**San Francisco**'   A fine plant for training as an espalier, and it also makes a fine weeping standard. The long tube and the sepals are rosy-carmine and the corolla is bright geranium-lake. The strong foliage adds to the appearance of this plant.

'**San Leandro**'   A strong upright grower with large double flowers freely produced. The tube and sepals are carmine and the corolla magenta verging on vermilion. In a large pot it makes a fine exhibition plant.

'**Sarong**'   Makes an excellent bush, with large foliage and a prolific show of bloom. The tube and sepals of the large double flowers are white, flushed pink, the sepals being long and twisting, while the corolla is violet-purple marbled with pink, the petals flaring wide open at maturity.

'**Scabieuse**'   A good variety for the house as it holds its flowers well indoors. Forms an upright bush with medium-sized blooms, the tube and sepals red and the corolla white, suffused blue on the outer petals and purple on the inner ones.

'**Shades of Space**'   A semi-trailer suitable for training as an espalier although it needs a shady position. The large double flowers are white on the tube and sepals, the latter pale pink on the underside, and the corolla is light purple fading to deep pink.

'**Shy Lady**'   A superb fuchsia that is quite easy to grow. Makes a bushy, upright and free-branching plant with large double flowers freely produced. The tube and sepals are ivory-white and the corolla is almost white with a tinge of pink.

'Sister Ginny' A fairly recent introduction of exquisite colouring. The medium-sized double flowers are pink on the tube and sepals, and the corolla is campanula-violet shading to fuchsia-purple. Of cascade habit the plant is ideal for a basket.

'Snowdrift' Almost a white self but with the corolla tinged pink. Forms a low bush with glossy dark green foliage against which the large flaring flowers show up well.

'Snowstorm' An icy greenish-white self with semi-double flowers on a plant of more or less trailing habit. Makes an unusual and striking basket plant.

'Sophisticated Lady' A lovely flower on a plant of a willowy habit that makes it ideal as an espalier. The tube and sepals are a very pale pink and the very full corolla is white. The double flowers are produced very freely.

'Star Light' An easy variety that is just about hardy enough to survive outside. The single flowers of medium size make a prolific show on an upright bushy plant. A very old variety with the sepals creamy-white and the corolla rosy-cerise.

'Strawberry Fizz' A fine trailer with large double blooms. The tube and sepals are light pink and the corolla a deeper shade of pink, with the petals picotee-edged.

'Strawberry Sundae' A cascade variety with a plentiful display of large double blooms, the tube and sepals ivory-white and the corolla strawberry-pink.

'Striker' Makes a good bush or standard of strong upright growth. A recent introduction. The tube and sepals of the semi-double flowers are flesh-pink and the corolla violet-purple.

'Sun-kissed' Naturally forms a strong upright bush but also makes a good standard. A vigorous easy grower with plenty of medium-sized double blooms. The tube and sepals are pale pink tipped with green and the corolla is pale red shading to crimson.

'Sunset' A fairly old variety that is still well worth growing for its long and prolific season of flowering. An easy fuchsia that may be grown as a bush, standard or trailer. The medium-sized single flowers are pale pink on the tube and sepals and orange-cerise on the wide-spreading corolla.

'Susan Travis' A fine hardy variety that makes a splendid summer bedder. Makes a medium-sized upright bush with a prolific display of single pink flowers, of a softer shade on the corolla.

'Sweet Leilani' The attractive colouring gives this variety a strong appeal, with the short tube and sepals of the double flowers a pale rose-pink and the wide-spread corolla a smoky-blue. Makes a strong upright bush with a profuse display of large blooms.

'Swiss Miss' A good basket plant with semi-double flowers. The tube and sepals are pale red, the corolla white with pink veining. The petals of the corolla spread out to give a skirt-like effect.

'S Wonderful' An upright grower with medium-sized double flowers. The tube and sepals are pink; the inner petals of the corolla are pale lavender, the outer ones orchid-pink.

'Tabu' A recent introduction of a free-flowering, upright habit. The tube and sepals are pale pink, the latter salmon-pink on the inside, and the corolla rose, marbled pale pink. Double flowers.

'Tammy' A beautiful double, with plenty of flowers on a vigorous spreading bush. The tube is waxy-white and the long wide sepals curving, twisting and upswept, a most attractive pink. The large wide-spread corolla is lavender, streaked with pink. Of vigorous habit it makes a fine bush or basket.

'Tangerine' Makes a strong upright plant with magnificent blooms well suited to the show bench. The fairly long flowers are freely produced. The colouring is unusual, with the tube flesh-pink, the sepals green and the corolla orange, aging to pink.

'Texas Longhorn' Not an easy plant but a magnificent one when well grown. The buds and flowers are enormous, the former often reaching 5in long opening to 7in across. The tube and sepals are bright red, the corolla pure white. Forms a lax bush needing much tying.

'The Aristocrat' A fine exhibition variety with large double flowers fairly freely produced on a strong upright plant. The tube and broad upcurved sepals are a pale rose and the serrated petals of the very full corolla creamy-white, veined with pink at the base.

'The Doctor' An easy grower both as a bush and as a standard. The tube and sepals of the single flowers are flesh-pink and the corolla rosy-salmon. A prolific bloomer with flowers of medium size.

'The Phoenix' A double variety with long flowers freely produced. The tube and sepals are rosy-pink and the corolla lilac. Makes a lax bush suitable for an espalier.

'The 13th Star' A very recent introduction with masses of bloom on a plant of trailing habit. The tube and sepals of the double flowers are neyron-rose and the corolla is of various shades of imperial purple.

'Three Cheers' A variety of unusual colouring. The tube and sepals are red and the corolla is palest pink at the centre changing to purple and blue at the edges. A striking plant for the greenhouse.

'Thunder Bird' A very free-flowering self-branching trailer, ideal for a basket. The large double flowers are mainly rose-pink and the corolla shades from vermilion to pink. Makes a good espalier.

'Tinkerbell' A plant of low spreading habit with long single flowers freely produced. The tube and sepals are pinkish-white and the corolla white, veined pale pink.

'Toby Bridger' A very popular variety of sturdy compact habit. The large double flowers are an attractive shade of pink and the similarly coloured corolla looks like a camellia bloom.

'Tom H. Oliver' A self-branching trailer with a profuse display of bloom. The tube and sepals of the double flowers are claret-rose and the corolla a dark ruby-red.

'Topper' Makes a strong upright bush with plenty of fairly large double flowers that show up well against the slender dark green foliage. The red tube and sepals contrast well with the dark blue corolla flushed with white.

'**Torch**' A fine exhibition variety with multi-coloured double blooms. The tube and broad sepals are a shining pink, the latter flushed with salmon on the underside, and the corolla is purple-red at the centre, with orange outer petals. Makes a tall upright plant.

'**Trade Winds**' A trailer with double flowers. The tube and sepals are a mixture of white and pink and the corolla is white with a touch of pink at the base of the petals.

'**Trail Blazer**' A popular basket fuchsia of vigorous cascading habit. The long tube and recurved sepals of the double flowers are magenta, with the corolla a slightly darker shade. Long, large flowers in profusion.

'**Trase**' A popular hardy fuchsia with small to medium flowers in profusion on an upright bushy plant. The tube and sepals are cerise, the corolla white with crimson veining.

'**Treasure**' Makes a good upright plant with plenty of large double flowers. The tube and sepals are ivory and pale rose respectively and the large fluffy corolla is violet-blue. A good exhibition plant.

'**Tresco**' A very hardy fuchsia of strong spreading habit. Although the flowers are small there are more than enough of them to make a good show. The tube and sepals of the single blooms are red and the corolla purple.

Tricolor' A form of the hardy *F. magellanica gracilis* with silvery pink-flushed foliage. The tube and sepals of the small single flowers are red and the corolla purple.

'**Trisha**' A large-flowered variety that can be grown as either a bush or trailer. The tube and sepals of the double flowers are almost white, tipped with green, while the corolla is burgundy flushed with rose and magenta.

'**Tropic Sunset**' Of willowy growth this should make a good espalier. The double flowers are small but plentiful and look well against the reddish-bronze foliage. The tube and sepals are carmine and the corolla dark purple.

'**Troubador**' A double-flowered trailer. The tube and sepals are a bright crimson and the large corolla is dark lilac-purple with splashes of crimson at the base of the petals.

'**Trumpeter**' A fairly modern *triphylla* hybrid. The terminal clusters of long tubular flowers are a pale geranium-lake. Makes an upright bushy plant with attractive bluish-green foliage.

'**Tumbling Waters**' A cascade variety with large flowers freely produced. The tube and sepals are a deep crimson and the corolla cyclamen-purple.

'**Ultramar**' A fine double variety with the tube and long, broad sepals a creamy-white and the globular corolla a delicate smoky-blue with some white-streaked petaloids at its base. Of strong upright growth it makes a good plant for exhibition.

'**Uncle Steve**' A free-flowering trailer with double blooms, pale pink on the tube and sepals and plum-coloured on the corolla. A good basket plant.

'**Victorian**' A double-flowered variety of strong upright habit; makes a good bush plant. The pink flowers are a self-coloured pink.

'**Violet Rosette**' An upright grower with bright green foliage against which the double flowers show up well. The tube and short broad sepals are carmine, with the latter turning straight back, and the large, very full corolla a deep violet with a touch of red at the base. The growth is upright and very bushy, with plenty of large blooms.

'**Vulcan**' A recent introduction forming a strong self-branching bush with semi-double flowers, the tube and sepals rose, tipped with green on the latter, and the corolla neyron-rose shading to ruby-red.

'**War Dance**' A semi-double with the tube and sepals white and the corolla purple, flecked with white and changing to smoky-orange and red. The lax growth makes it suitable for either bush or basket culture.

'**Whirlaway**' A semi-double self of a creamy-white, tinted pale pink with age. The short tube and long sepals give the bloom an unusual appearance and the long flowers are freely produced on a plant of willowy habit. A good exhibition variety.

'White Galore'   A double-flowered trailer with large blooms, pure white on the tube and sepals and near-white on the corolla. A good basket plant.

'White Pixie'   A near-hardy variety of upright compact growth, with a generous display of small single blooms. The scarlet tube and sepals and the white corolla make an effective show against the yellow-green foliage.

'Wild and Beautiful'   A very recent introduction with double blooms on a free-flowering plant of semi-trailing habit. The tube and sepals are neyron rose and the corolla is dark amethyst-violet.

'Wings of Song'   A cascade variety with the tube and sepals pink and the corolla lavender-pink. The large flowers make a particularly attractive basket.

'Winston Churchill'   A good double of attractive colouring, with the tube and sepals pink and the corolla silvery-blue. A fast grower with a long flowering season. The medium-sized flowers are produced freely on a dwarf, very bushy plant.

'W. P. Wood'   A hardy fuchsia that also makes a good summer bedder. The tube and sepals of the single blooms are a rich deep red and the corolla is royal purple, paler at the tip of the petals. The flowers are produced freely on a dwarf, very bushy plant.

'Yankee Clipper'   An upright grower with double flowers. The tube and sepals are carmine and the corolla ruby-red variegated with carmine.

'Yonder Blue'   Makes a strong upright bush with large flowers fairly freely produced. The tube and sepals are rosy-red and the corolla dark blue.

'Ziegfield Girl'   Makes a lax bush with plenty of medium-sized blooms. The tube and sepals of the very double flowers are a very deep pink and the corolla is a slightly lighter shade.

Suppliers

Great Britain

H. A. Brown, 20 Chingford Mount Road, South Chingford, London E4

B. & D. Davies, 2 Wirral View, Connah's Quay, Deeside, Clwyd

Deanburn Fuchsias, 30 Dean Place, Seafield, Bathgate, West Lothian

Deeside Gardens, (J. Penhall), 37 Grains Road, Shaw, Oldham, Lancs

The Fuchsia Nurseries, (R. & J. Pacey), Strathern, Melton Mowbray, Leics

Fuchsiavale Nurseries, Stanklyn Lane, Summerfield, Kidderminster, Worcs

High Tree Nurseries, Buckland, Reigate, Surrey

Hillcrest Nurseries, (C. J. Castle & Son), Church Bank, Goostrey, Nr Crewe, Cheshire

Hillcrest Nurseries, Ratcliffe Highway, Hoo, Rochester, Kent

D. T. & J. A. Hobson, 130 Aughton Road, Swallownest, Sheffield

Homestead Nurseries, (W. A. Shipp), Oxford Road, Denham, Uxbridge, Middx

Jackson's Nurseries, Clifton Campville, Nr Tamworth, Staffs

J. E. Knapp, 21 Jubilee Avenue, East Harling, Norwich

Landau, Promises, 8 Stanley Road, Peacehaven, E. Sussex

C. S. Lockyer, Landsbury, 70 Henfield Road, Coalpit Heath, Bristol

A. B. Longden, 9 Middleton Road, Clifton, Rotherham, S. Yorks

Markham Grange Nursery, (V. J. Nuttall), Brodsworth, Nr Doncaster

Mounts, Waterloo Road, Radstock, Nr Bath, Avon

S. & E. Orton, 6 The Castleway, Willington, Derby

James Travis, 394 Brindle Road, Bamber Bridge, Preston, Lancs

The Vernon Geranium Nursery, (Janet & Derek James), Hill End Road, Harefield, Uxbridge, Middx

Ward Fuchsias, 5 Pollen Close, Sale, Cheshire

R. Warren, 307 Highbridge Road, Sutton Coldfield, W. Midlands

Woodbridge Nursery, (Don Stilwell), Hibernia Road, Hounslow, Middx

United States

Airport Nursery (Richard Carr), 42867 Airport Rd, PO Box F, Little River, CA 95456

Andy Cracolice's Fuchsia Garden (Andy Cracolice), 1088 E Campbell Ave, Campbell, CA 95008

Annabelle's Fuchsia Gardens (Regine & Bruce Plows), 32531 Rhoda Lane, Fort Bragg, CA 95437

Barbara's World of Flowers (Barbara L. Schneider), 3774 Vineyard Ave, Oxnard, CA 93030

Berkeley Horticultural Nursery (George Budgen & Ken Doty), 1310 McGee Ave, Berkeley, CA 94703

Bob Thomsen's Garden Center (John & Iris Watson), 1113 Lincoln Ave, Alameda, CA 94501

Burns Nursery & Greenhouse (Gean W. Burns), 2104 North 250 West, Sunset, Utah 84015

Chipmunk Hill Nursery (Mrs Phillip E. Hester), PO Box WW, Port Orchard, WA 98366

Christy's Fuchsia Farm (Christy Nelson), PO Box 171, Mount Eden, CA 94557

Cotton Candy Nursery (Max & Maxine Scholz), PO Box 204, Cutten, CA 95534

Dale's Flower Nursery (Dale & Dorothy Taylor), 474 North 54th St, Springfield, OR 97477

Dayka Nursery (William Dayka), 1643 Calle Canon, Santa Barbara, CA 93101

Dott's Greenhouse (Leona Emburg), 11560 SE Stark St, Portland, OR 97216

Elk Valley Azalea and Fuchsia Gardens (Winona B. Anderson), 2901 Elk Valley Cross Rd, Crescent City, CA 95531

Empire Nursery (Bob Cowles), 3747 Guerneville Rd, Santa Rosa, CA 95401

Emrick's Fuchsias & Begonias (Nola & Martin Emrick), 7303 Renton-Issaquah Rd, Issaquah, WA 98027

Faye's Fern & Fuchsia Gardens (Robert Faye), 31951 Highway 20, Fort Bragg, CA 95437

Fire Mountain Fuchsias (Annabelle Stubbs), 1182 Avocado Rd, Oceanside, CA 92054

Forest Glen Nursery (Thomas F. Lowney), 1000 River Rd, Fulton, CA 95439

Fuchsia Forest Nursery (Robert A. Castro), 9234 East St, Oakland, CA 94603

Fuchsia Odyssey Nursery & Landscaping Service (Paul & Nicole Savage), 427 31st Ave, San Francisco, CA 94121

Fuchsiarama (William Barnes & Cliff Ebeling), 23201 North Highway 1, Fort Bragg, CA 95437

Fuller Hill Nursery (Jeanette M. Pettit), 413 Keyes Rd, Elma, WA 98541

Guasco Fuchsia Gardens (Curtis & Mary Updegraff), 515 Aspen Rd, Bolinas, CA 94924

Half Moon Bay Nursery (Ron Mickelsen), Highway 92, Half Moon Bay, CA 94019

Hassett Fuchsia Garden (Chuck & Mary Hassett), 3413 Nevada St, Eureka, CA 95501

Hastings Fuchsia Gardens (Mrs Robert W. Hastings), 302 South Nardo Ave, Solano Beach, CA 92705

Betty Herrier, 175 Oak Road, Danville, CA 94526

Hidden Springs Nursery (Hector Black), Rt 14, Box 159, Cookeville, TN 38501

L & J Nursery (Lem & Jackie Downs), 1690 East Fork Rd, Williams, OR 97544

Minkler's Nursery (Jean & Robert Minkler), 2218 Callendar Rd, Arroyo Grande, CA 93420

Nix Nursery (Nicholas Ostopkevich), 4707 Cherryvale Ave, Soquel, CA 95073

Ocean Bluff Fuchsia Gardens (Frank & Francesca Stasko), 33625 Pacific Way, Fort Bragg, CA 95437

Ray's Fuchsias (Ray & Vivian Schmidt), 6240 Goddess Ct, San Jose, CA 95129

Redwood Grove Nursery (Gary von Nostitz), 8306 Bodega Ave, Sebastopol, CA 95472

Rhea's Fuchsia Garden (Rhea M. Beardsley), Rt 2, Box 64D, Burton, WA 98013

Robert G. Eckel Greenhouses (Robert G. Eckel), PO Box 12, Brooklin, Ontario, Canada LOB 1CO

Robert Morris Fuchsias (Robert Morris), 2461 West Avenue 134th, San Leandro, CA 94577

Shadow of the Mountain Nursery (Dolores M. Short & Erik R. Wilk), 22522 South Fellows Rd, Beavercreek, OR 97004

Sherry's Fuchsia Gardens (Sherry Klonius), Smith Creek Rd, Rt 2, Box 350, Raymond, WA 98577

Silver Sun, Inc (J. S. Berglund), PO Box 307, Cashiers, NC 28717

Stubbs Fuchsia Nursery (Robert L. Meyer), 737 Orpheus Ave, Leucadia, CA 92024

Three Firs Nursery (Dickie & Winnie Lofton), Rt 1, Box 289, Amity, OR 97101

Valley Garden Center & Florist (Mrs Irma E. Smith), 2646 Santa Maria Way, Santa Maria, CA 93454

Verona's Plants (Mrs Verona Calvert), 93490 Prairie Rd, Junction City, OR 97448

Western Garden Nursery (Ron Marciel), 28191 Hesperian Blvd, Hayward, CA 94545

Join a Fuchsia Society!

In most areas of Britain there is a local fuchsia society but as the secretaries of these are liable to change at any time, start by contacting the secretary of The British Fuchsia Society, at 29 Princes Crescent, Dollar, Clackmannan-shire, FK14 7BW, Scotland. He will be able to advise you on your nearest local society.

In the USA, contact The National Fuchsia Society, c/o Southwest Botanic Garden, 26300 Crenshaw Blvd, Palos Verdes Peninsula, CA 90274, or the American Fuchsia Society, Hall of Flowers, 9th Avenue & Lincoln Way, San Francisco, CA 94122.

General Index

Index of Varieties